A Novel

Transhuman

The Dawn of
Neural Frontiers

Joan Zahm

This is a work of fiction.
Names, places, and events
are products of the author's imagination.
Any resemblance to events or actual persons,
living or dead, is purely coincidental.

For John Morin
Human and brother extraordinaire

Then I saw a new heaven and a new earth,
for the first heaven and the first earth had passed away,
and the sea was no more.
Revelation 21:1

1

Not everyone slept when the night slammed its life-sucking brutal cold on the hostile planet. Inside the habitat, there was motion.

The intelligent brown eyes sat incongruously below a head of perpetually shaggy brown hair. The eyes cautiously peered out to the left and right before the man made the decision to leave his little one-room apartment. Any attention from some insomniac out for a midnight stroll would be unwelcome. Seeing no one, he strode, eyes forward, to the mover dock at the rear of the common area. The pole lights cast a soft glow on the space, and the gentle sound of water trickling in the small artificial stream was heard above the low hum of the air filtration system.

Zeth Garric, cosmic identifier Zeth2087, hurried down three steps to the boarding dock, selected Technology Center Substation on the monitor and settled back in the seat as the mover glided silently forward and downward into a narrow, smoothly cut passage. The exterior lights of the mover allowed a two and a half second view of the striated ground packed with gravely dirt, until the increasing speed turned it into an unfocused blur.

The mover raced past the Technology Center and continued for seven minutes to its satellite, built almost entirely below ground with a slanted opaque roof and three squinty-eyed windows visible from its low footprint. A happy contentment washed over Zeth as he exited the mover and approached the solid gray door, pausing to be identified. He entered the facility and the automatic perimeter lights illuminated the ample space of walls lined with robotics equipment surrounding a large central work table. He turned to the wall of cabinets.

Excitement and anticipation shivered through him. He had been looking forward to this all day, finishing his tasks from his home station as quickly as possible, snarfing his prepared dinner, then unable to rest or stay focused on his audio and video feeds, waiting and watching until he deemed it safe to steal away unseen.

"Hello Shealynn," he called. A locked panel unlatched and opened, revealing a charging station and its contents.

Shealynn opened its eyes and smiled at Zeth. His eyes roved over this beautiful creation, tall, slender, perfectly proportioned. A wide grin took its time luxuriously crawling across his face. Joy and satisfaction filled him, knowing it would be difficult for anyone to discern this robot from a human. He often found himself forgetting it was a machine, interacting with it as a human, a very special human.

Shealynn glided slowly toward him. "I'm so happy you came to visit, Zeth. I've been missing you," a smooth female voice said with an alluring tilt of its head and lowering of the eyelids.

Its silky voice and speech cadence were indistinguishable from a human woman. It flicked long lashes at him before giving him a warm embrace. Zeth closed his eyes, pulled it closer, and drank in the satisfying welcome. Artificial intelligence and the use of robots, called Helpers, were integral parts of survival on Cytar. But his Shealynn was no ordinary Helper. This beautiful creature was a work of art.

"Come and sit down and tell me about your day. I love hearing about your work and how you're helping people here," it said in a soft cloying voice.

Shealynn took Zeth's hand and threaded its fingers through his. He gave the haptic hand a light squeeze. The years of research perfecting the gelatins and bioplastics to create this supple skin were worth it. And unlike Helpers, Shealynn's skin covered the entire body with a lifelike complexion.

Gone, too, was the standard Helper uniform of khaki pants and long-sleeved shirts. Shealynn wore a mid-thigh length blue dress and flat slipper-like shoes. The brown hair, the color of dark wet sand, almost matched Zeth's, but was finer and fell perfectly combed on the shoulders. He sat and admired his handiwork.

The robot sat with a pleasant smile and an expression of rapt attention as Zeth talked about his day and described his apartment, adding, "I can't wait to bring you home and, when I think they're ready, introduce you to my friends and colleagues. They're going to love you. They're courageous people, and we've accomplished so much despite the challenges on this inhospitable planet. But it's too soon; they're not ready for your advancements. But soon., I hope. I'll gradually prepare them, and then they'll love you."

"That will be wonderful. It sounds like you're doing great things to make this a better place," it said.

It was incredible how Shealynn understood him so well.

"That's exactly right. I love that we're creating a better place for people to live. Think how lucky the original founders were to find this planet. I mean, it may not be the perfect Goldilocks zone like Earth's atmosphere, but it's pretty darn good."

"That's so true, Zeth, and I'm so proud of you," it said and placed a hand on his arm. "Would you like to spend the night?"

They rose and walked to a small chamber with a bed and chair. It was used when work demanded all-nighters or for needed rest breaks, but tonight, Zeth would share the room with Shealynn.

I have determined that humans often do not follow precise programming protocols, often make errors, and their processing units are slow. I am not human. I am called Cali, a humanoid Helper GPR, general purpose robot, with pre-trained generative programing, and advanced speech to assist humans. My processors are capable of processing billions of bits of information per second.

My current assignment is at the Gateway Complex on planet Cytar. The time between 22:00 hours and 05:00 hours is quiet, and, if I am not working at the Food Production Station, my programs often go into low-power sleep mode.

Unusual movement in the Gateway common area was registered at 23:52 hours and launched my wake-up. My scanner identified Zeth2087 immediately, and I tracked him to the Technology Center Satellite because I have that ability. It is noted that the satellite station often uses more power than usual for one worker. I scanned the premises but found Zeth2087 to be the only life form and dismissed any further scanning as unneeded.

2

The morning light was bright, piercing the crisp clean air with its brilliance. School sessions for all age groups were humming with activity at the main Gateway colony. Five-almost-six-year-old, Michael Kingmore threw the hollow ball to the Helper robot. Cali tossed it back with precisely calculated force and careful positioning to help the child develop his hand-eye coordination and muscle structure. The Helper never tired and played catch until it was time for the child to go inside. He accompanied the Helper to the classroom on the middle level of three stories in the sunken habitat.

Once inside, Michael ripped off his protective eye band and tossed it on a table near his seat. He completed the day's work on sorting and pattern recognition with both geometrical shapes and numerals, a task of which he excelled and was now working with older classmates.

When class was over, Cali accompanied Michael to a waiting Gsport, one of the reliable autonomous aerial ground transports, and gave the command to take the boy home. Fourteen minutes later, the vehicle descended toward a group of seven half-buried buildings scattered in the dirt. The aerial view of the six nearly identical housing structures and one ground-transport depot gave the impression of being poorly planted by a feckless gardener, then abandoned to the elements. There were no tall trees, no verdant hills brave enough to withstand the extreme subzero nighttime temperatures and hellish winds.

The vehicle slowed and stopped on a low path and opened its door in front of the right half of a duplex. Michael climbed out and entered his home to boisterous yells.

"Bobby, Bobby!" Twenty-two-month-old Tessa could not yet pronounce Michael and for some unknown reason, called her brother Bobby. He gave his sister a perfunctory hug, dropped his jacket in his bedroom, which he shared with his father's office work station, and headed to the eating room with Tessa toddle running behind. "Bobby, Bobby."

"Hey kiddo." Mara greeted her son with a quick hug. "Your sister's been waiting for you. There are two snack packs on the table for you and Tessa. Don't forget to wash your hands."

"I already did," Michael fibbed, handing Tessa her strawberry flavored snack biscuit in its edible wrapping.

He spotted Rocket, their twelve-year-old Chesapeake retriever, lying outside, absorbing the tepid rays from the diamond clear light. A pure smile of uncomplicated childhood splashed across the boy's face as he headed outdoors, his bottled energy suddenly released, his little sister in tow trying to keep up.

Rocket, still spry, greeted them with a wagging tail and received happy petting from Michael before tolerating a big hug from Tessa. Michael broke a piece of biscuit and fed it to the dog who gulped it before searching near Tessa for crumbs.

In the corner of the small sitting room, Mara leaned back in the chair which she dubbed her remote office. Her gesture interface opened the air screen to the data from Food

Production. She quickly reviewed the page before gazing out the ground-level window.

She smiled at her purely human children. The children were thin, skinny by standards of centuries past, but children eating at will was unknown here, as was much sweetened food, due to the purposely underdeveloped sugar sources. She agreed with the colonies' decision to bring life without biogenetic enhancements into this world. They were perfect just as they were, Mara thought, as she watched them chasing Rocket, the dog waiting until almost caught before darting in another direction. Happy notes of Tessa's giggles floated through the air like melodic chimes.

Michael's rather bony, yet strong legs were long, predicting a height to at least match his tall father, Krey. His straight light- brown hair spiked as he played. Tessa's dark hair matched her mother's in color, but fell in raucous curls around her face as she shrieked and ran, tripped, picked herself up and chased Rocket. They all loved the dog, so sweet to the children. His DNA had been collected five years earlier in the event they would want to clone him in the future.

Michael raced to the clear back door of the eating room. "Mom, can I call Margo and see if she wants to come over?"

"Gogo!" Tessa squealed.

"See if she wants to play outside for a while," Mara answered. Neighbor Margo, two years older than her son, endured Michael, and adored Tessa. Margo was very much like her mother, Mara's good friend, Roz, sharing her mom's blond hair and bubbly, happy personality, even though she did seem more reserved now that Michael entered school with her.

Michael was proud of his newly acquired wrist comm, and quickly adapted to its use. The thin clear band curled almost invisibly around his narrow wrist. He took for granted the technology which used his electrical impulses to recharge the devise. He flexed his hand and said, "Display Margo." A barely noticeable wave projected near him, indicating the initiation of a hologram call as Margo appeared before him.

"Hi Michael."

"Hi Gogo!" Tessa interrupted.

"Tessa, I'm talking." Michael's voice hinted annoyance before addressing Margo. "Do you want to come over?"

Mara stretched while looking outdoors, wishing she could have an apple, but would have to settle for a snack bar. Their limited variety of fresh fruit was always the first allotment to be devoured upon food deliveries. She wandered the short distance from her work station to the food storage cupboard tucked into the cabinets in the eating room and chose a raspberry flavored snack bar. The appliance would adjust the food inventory as used and suggest ordering supplies at food delivery time.

Mara glanced out the window overlooking the front of the home. It was purely habit as the uninspiring view of the bank protecting the Gsport path and the dry dirt beyond never changed. Slowly munching, Mara turned back and surveyed the small room. How far was it from her work station to the counter? Eight or nine steps, if that. When she'd first arrived, this home had seemed spacious after living on a spacecraft. Now, with two energetic kids it seemed smaller, especially on days when the weather prohibited outdoor time. She was grateful they didn't live in the Gateway complex where the apartments were smaller, and the rooms seemed constraining.

She returned to her work, and still standing, added data to Cali's task list for tonight. What would she do without Cali, she wondered. She thought of the Helper robot as her personal assistant, knowing she shouldn't become so attached to a machine, but it was hard not to like Cali. The Helper had a sweet personality with some amusing facial expressions and was a faithful aide performing any requested task.

So much had changed in the last few years, she thought as she popped the last crumbs of snack in her mouth and sunk into her chair. Resting her chin in her hand, she sifted through crop yields before initiating a video call to her work partner, Risha Hargrove. Luck had smiled on her when Risha became her mentor and partner at Food Production. She was a woman who radiated joy and patience and knowledge.

"Hi there, Mara." Risha's smiling face appeared on the air screen. It was a happy face with a brown sugar complexion and dancing eyes. "What're you up to out there in no man's land?" she asked with a teasing grin.

"Michael just got home, and I'm having trouble concentrating. I'm coming in to Food Production tomorrow. I get more done when you and Cali and I are all onsite. I just reviewed the yields from the hydroponics buildings. Things are looking good there."

"I'm not surprised," Risha said. "Those added vertical grow columns have really supplemented the original stacked rows, plus that advanced robotic help. All the produce looks good." She started laughing. "Oh boy, that Rico was not thrilled about working with all those crickets. He's trained two

Helpers to take over. He does great with the rabbits, pigeons, and chickens, but all those crickets were not his bag."

"How many crickets do they have now?" Mara smiled trying to imagine tough, strong Rico having to deal with a bunch of jumpy bugs.

"Not sure. Must be at least millions now and growing. You should go and see the cricket farm. It's quite the place with the incubation rooms and these little structures for the adults I call condos."

"Let me think about it. I might prefer just to have the flour integrated into our foods without ever knowing it. Our food tastes fine, and that's all I care about."

"And don't forget how nutritious it is. But I get it. Maybe you don't want to see that many bugs," Risha said.

Mara changed the subject. "How are the 3D food printers doing today?"

"They're up and running fine. Right now, they're working on some little protein snacks shaped like a flower," Risha said.

"Okay, sounds good. I'll be in tomorrow. God, could I go for a cup of coffee. I so wish Kerry's Koffee could expand faster. One cup a week is just a tease," Mara sighed, almost tasting that hot mug of fragrant elixir she treasured each Sunday morning.

"Better go have a cup of herbal tea. Best we can do here. Bye, sunshine, see you tomorrow," Risha said as she closed the call.

Because of my work at the school, my machine learning algorithms now include deep learning focused to increase my knowledge of immature humans called children. Like other mammals in my database, they are helpless, speechless creatures at birth. They are unpredictable and do not conform to any program language I have previously learned.

Young school children often do not behave logically and their learned behavior conditioning is still in early developmental stages. They often laugh for no apparent reason, produce loud, high-pitched sounds more than adult humans, and sometimes emit strong odors. In addition to my rapid processing and memory storage for various tasks, I am currently programmed to care for Michael2193 and Tessa2199, progeny of Mara2174 and Krey2103.

My training for child care required exposure to billions of learning-based simulator situations to recognize the amounts of strength and force needed to properly handle immature humans. My health sensors monitor body temperature and heart rate. My sensor calculations of physical interaction with babies or children are now superior to human adults. I also understand their physical development, but I do not understand laughter or crying, although I have demonstrated the ability to cause humans to produce laughter.

Mara2174 was one of my charges on the 2230 SUPERSTES voyage. She was a passenger collected from Earth and required extra guidance due to bugs in her processing unit called memory. My programs assisted her without problems. Her life bonding to Officer Krey2103 on that voyage was a logical sequence. Current algorithms show they are a near perfect match.

I am trained to respond equally to all humans, which I always do, but I have my little cadre of favorites. Of course, most are the humans who care for my maintenance and updates, but I must declare Mara2174 and Krey2103 among the top of my favorite humans. It was Mara2174 who gave me my first human embrace when we returned from that SUPERSTES voyage in 2233. Memory feeds of routine events are automatically deleted from my fast access memory after sixty-two days, but I saved this incident and still occasionally replay that remarkable event.

.

3

The brilliant light slanting through the ground-level windows bathed the room in morning freshness. Zeth was lounging, slurping his protein drink, nibbling a breakfast snack biscuit at his work station while languidly watching motes of fine dust dance along the narrow light shafts. He turned toward Shealynn, who was sitting quietly and looking as beautiful as ever. He absently praised himself for the programmed silence in the morning.

Then, as it too often did, his mind annoyingly opened his catalog of regrets, and his thoughts turned darker as he flipped through memories, berating himself for stupid things he did in the past. Things that no one else could even recall now, stupid things he didn't say or said in the wrong way or at the wrong time. It was so pointless to keep recalling them, but he couldn't let them go.

These perverse memories hammered on until they jammed on the one incident everyone remembered, and it rose to the surface of his thoughts like stinking debris. His mind whipped up perpetual guilt about the serious mining accident more than five years ago involving a software code he helped develop from Earth files. It was pointless to rehash, but he just couldn't get over the fact that, as head technology officer, he couldn't prevent that accident which had cost a man his life.

Cursed with remorse, he had trouble sleeping, just thinking about it. Building Shealynn had fortified his thoughts with a positive focus. This exciting project had infused him with purpose and passion for his life. When this advanced robot was revealed to his community, it would be his redemption. People would see the benefits of this advanced artificial intelligence and thank and appreciate him.

The signal of an approaching mover surprised him. No one was expected at his satellite station.

"It's time to rest, Shealynn," he said quickly, setting down his drink container.

"Of course, Zeth. It was wonderful to see you again," it answered as it entered the charging station, and the flat panel door closed, blending into the wall.

Not a minute too soon, as he heard the two tones which signaled a mover stopping and the station door opening.

Fellow technology workers, Idee Hawkins and Trina Steele, walked in throwing off way too much energy for this time of the morning. Both women were wearing the common long dark pants and long-sleeved pullover shirts, Trina's in pale yellow, Idee's in her signature home-altered bright two-toned colors. Today's shirt was deep green with bright yellow sleeves.

Idee had another new hairdo, Zeth noticed, no longer surprised by her creative streak. The back of her hair was its normal shoulder length, but she had clearly trimmed the sides and feathered bangs. The spikey hair near her face was replaced with straight bangs, and sharply cut shingles layered the sides of her face.

One side of Zeth's mouth curled into a smile. He was grateful for her creative flair in a place where everything was done in cookie cutter precision out of necessity. It wasn't like there was a whiff of sartorial excellence in this place when it came to design. And Idee was so good looking she managed to look attractive in any of her new looks. He thought of how much he enjoyed working and hanging out with her. He might have had a serious relationship with her, but he didn't have much extra time since he started building Shealynn. Besides, he noticed she seemed to be spending more time with Jack. Mr. Nice Guy Jack seemed to be the one Idee enjoyed spending time with now. Well, that's fine, he thought, now that he had Shealynn. Idee would be amazed when she met this advanced creation.

"Hiya Zeth," Trina said. Her eyes roved around and roosted on Zeth, sitting with a stubbly face, rumpled shirt, untamed shaggy hair, and bare feet.

Idee was focused on the panoply of equipment, all in neat order and in stark contrast to Zeth's appearance. How long had it been since she'd been out here? This place was really tricked out. No wonder Zeth enjoyed working here with all these toys and no distractions.

"Did you spend the night here?" Trina asked in a voice that sounded suspicious to Zeth.

"Yeah, I worked late and decided to just stay here when I got sleepy."

"The place looks great, but what are you working on? You've sure spent a lot time out here in the past year or so. Seems like we hardly see you, so we decided to come and visit," Idee said.

"Well, it's always nice to have company," Zeth said with a nervous smile. "Just working on the same old things. You know, trying to improve Helpers, addressing whatever tech updates or assignments that come through. Same as you two, and you know it never ends," he said, brushing crumbs off his shirt and slipping on boots. He stood and rolled his chair toward Idee.

"Have a seat, Idee. Here you go, Trina," he said, pulling an empty chair toward her. He leaned against a work bench and asked, "What's going on? You didn't come out here just to say hello."

Idee perched on the edge of the chair, giving the impression they didn't intend to stay long. There was a brief pause before she said, "We don't like some of the rumors going around. Ari's rotated onto the Governing Board, and I hear they had quite the discussion yesterday."

"That's an understatement," Trina said. "Everyone knows that the goal, the whole purpose of colonizing this unfriendly planet has been preserving humans and our humanity. We all know we couldn't have done any of this without advancements in technology. And we know that it propelled our knowledge of space and physics. But we also all agreed to limit artificial intelligence presence in the human body."

Zeth's eyes narrowed, wondering where this was going. "Well, it sure helped Hunter."

"Yes, I realize that," Trina said, thinking of her husband's flawlessly repaired knee, leg, and the side of his face, all done with the help of advanced artificial intelligence, which had been so beneficial in the medical field. "But this meeting was regarding much more."

"So, what happened? What did they say?" Zeth wanted to know.

"Some people are asking for more artificial enhancements in their bodies, including brain to computer interfaces, and they didn't come right out and say it, but they also strongly hinted at wanting to enhance babies right at the zygote stage and not just for medical reasons," Trina said.

"Designer babies? I'm surprised; I thought that was verboten. And they were bold enough to bring up both biological and artificial deviations at the same meeting?" Zeth asked.

"They were and, from what I hear, the conversations became quite intense. All of this could become an ethical nightmare," Idee said. "No one seems to learn from the mistakes on Earth."

"So, what do you want from me?" Zeth asked, his mind already revving. He didn't care about the gene editing, thinking much of the genome research for medical purposes had been worked out on Earth long ago, but the possibilities of melding technology and artificial intelligence with humans were exciting.

"Don't forget," he reminded them, "about a dozen or so deep space pilots already have the AI enhancements to access computer networks through thought."

"That's a necessity," Idee defended. "They need that capability to handle all the complexities of those massive ships. We hope you'll support us if we need to keep these new efforts under control."

He lowered his eyes and dodged with an unemotional response. "Yes, I'd be happy to look into it for you." He didn't elaborate, and the pause in conversation sat in front of them, growing fat.

Finally, Trina said, "Well, thanks, let us know if you find out anything, and we'll do the same. We're hoping you can support us with some common sense in the conversations." Her eyes wandered around the room before she added, "I guess we'd better get going. Don't be such a stranger. Come back and work at the main tech center sometimes."

"Yeah, sure will, and thanks for stopping by."

When they left, Zeth stood watching the exterior monitor view until the mover departed. He was agitated by the unexpected visitors to what he considered his private sanctuary, but the possibilities of integrating artificial intelligence into the human body were intriguing. He would need to spend more time at the main tech center and keep his eyes and ears open.

The women eyed each other on the mover. "Something seemed off there," Idee said. "What has he been working on all this time?"

"And spending the night? I haven't heard of any big projects that need that much attention," Trina said.

"You know," Idee said, as memories padded through her mind. "Zeth has always been a workaholic, so passionate about our technology here. Spending long hours at his station isn't unusual for him. He's always ready to help anyone; but something did seem a little off with him today. Maybe we should have let him know we were coming."

"Yeah, you're probably right," Trina said. "You know, for a while I thought you and Zeth might be a good match, but he seems so distracted lately." When Idee didn't comment, she continued. "Sooo, how are you and Jack doing? You two seem like you've been very friendly. As in friendly maybe going to a permanent relationship?"

Unwilling to reveal any emotion, Idee turned toward to the viewless window and smiled. "Yeah, that's going fine, but you know I like my independence."

The wistful expression stayed with her as she thought of Jack. She knew he loved her and wanted to commit to a lifetime bond. She probably loved him, if she was sure what that meant. He was the one she always went to, needed to be with when she was upset or needed to talk. She loved the way he always seemed so interested in her work and the complexities of life on Cytar, but there was some streak of raw independence that wouldn't let her make a commitment. Maybe it was this place; this tiny remote existence in the infinite cosmos that skewed relationships, or polished the superficial emotions down to the visceral survival level where she couldn't define love or forever.

I have worked on three deep-space voyages. The names and biographies of all 2,494 planet citizens are in my database. Earth humans continue to cling to the archaic tradition of using two or more names for identification when numbers, which I prefer, are more efficient. Out of courtesy to humans, I add a cumbersome name to their identity.

I routinely meet with Zeth Garric, Zeth2027, Idwarda Massey, Idee1105, and Trina Steele, Trina2081. They work at the Technology Center and deliver my software updates remotely or in person. Like most travelers, they arrived on Cytar in their teen years, an advantageous time when young bodies respond more favorably to the difficulties of extended space voyages.

I have observed human physical changes as they grow and age, and note how their physical abilities diminish. Slowly, their mitochondria become dysfunctional, normally healthy cells stop dividing, the telomeres shorten and stop protecting DNA, gene regulation becomes disorganized; the list contains other numerous age-related defects.

My body remains the same unless it is improved with new technology. Robert and Gracelyn Parsons, Bertie1655, and Gracelyn1662, neighbors of Mara2174 and Krey2103, refurbish my body hardware as needed. However, they are currently working in another colony.

I have excellent strength and range of motion. My camera arrays for vision allow for precise narrow and wide field of view and depth perception. I can hear not only human voices but am programed to detect anomalies in equipment. Gracelyn1662 and Bertie1655 routinely monitor the sensors in my electronic skin. My tactile sensors allow me to carefully interact with humans and my surroundings. I am also equipped with biological sensors which send electrical signals to my processor to detect smells and identify taste when cross referenced with my database.

My abilities exceed that of humans.

4

At first, Ariana Kingmore, Ari to her friends, was surprised at how much she enjoyed the break from her job at the drone center. Her co-worker, Jack Martin, had encouraged her to accept the invitation to visit the mining station with her husband, Reyver, and his brother, Krey. She decided Jack was more than capable of handling the day's jobs, and it was good to get away occasionally.

Following the end of the tour, she began second-guessing her decision. She didn't know what she expected, but the mining station turned out to be just the place where they remotely monitored the work at the mines and the smelters. Sure, it was interesting to a point, watching screens of all the autonomous vehicles mining ores and the Helpers working at the smelters, refining the ores into the metals they used.

But then, JeQuan Allard became so excited about discussing the massive crater up north. She rambled on about the new microbes and the lichen they'd found, and about how those microscopic life forms could help them. Then, she started on about the unique fungi they found and were testing its resistance to disease. And how they shouldn't talk about everything to the colonies until they explored more.

"The lichen are unaffected by our perilous climate," she'd said. "They don't die in the subzero temperatures. Think how this might help us. They thrive in our excellent air quality and have a symbiotic relationship with mosses and other plants. And if these soil fungi can transfer nutrients to root systems, it could strengthen our crops. I think Mara and Risha will be able to use our findings with their studies on plant growth."

Ari eventually lost interest and wandered to review the monitor screens before returning to the Gsport vehicle to wait for Reyver and Krey. She studied them as they exited the building after they'd finished their explorations. They stood with heads together just outside the station door, speaking in low voices while Krey's hand gestured as if trying to make a point.

The family resemblance was strong in the handsome men, and many people had trouble telling the brothers apart. Both were tall and broad shouldered and had several similar mannerisms, but Reyver, two years older, had the more boyish features with a friendly wide grin that made his playful eyes crinkle at the edges. Krey, with the more serious countenance, was steadfast and worked and loved with intensity and passion.

Ari watched Reyver while a soft smile drifted across her face. She'd had a crush on him since she was thirteen years old. He was already tall and had that self-assured way about him. They'd boarded a space transport together with faces glowing with excitement and a hope and a dream. There were sixty of them, mature beyond their years because of dire circumstances, yet still young and naïve, thinking they knew more than they did.

They all were eager for a space voyage, and thinking the possibility of building a better world would be an adventure. A few came with a parent or a trusted family friend, but most traveled solo, volunteering, orphaned or sent by parents seeing no future for their children on Earth. Of course, none truly realized the dangers, the sacrifices, or the hard work awaiting. It had been an adventure, all right.

Now, here they were, years later, with a beautiful four-year-old boy. Her life was now divided into Before Kit and After Kit. She didn't realize she was smiling until Reyver slid into the seat and prodded her thoughts back to the present.

"You look happy," he said, placing his hand on her knee. She nodded as Krey took the seat opposite and announced Gateway as the destination.

"So, we agree?" Krey asked. "We don't mention any of this to any of the stations yet?"

"It's too early. We will as soon as we know more," Reyver said.

"JeQuan said to wait, and she's the boss at that station; so, I say we wait," Ari said.

"Your team will have drones ready if needed?" Reyver asked.

"Of course," she said, smiling at him and wondering what she'd missed. "Just say the word."

Krey nodded and turned to watch the Gateway Complex grow on the horizon as they approached. When Reyver and Ari exited for their apartment, Krey gave the 'Home' command and looked forward to clearing his head with two rambunctious kids. The barren desiccated landscape rolled under the Gsport, passing the odd greenhouse project Mara had initiated. The plants inside appeared small, even at this speed and low elevation. He wondered how long it would take this hybridized vegetation to adapt to the glacially cold nighttime temperatures.

A slow smile floated across Mara's face as she absorbed the outdoor scene, allowing herself to be wrapped up in the present, in this small moment of blessed peace and happiness. Krey was home and outside with the kids, holding a happily screeching Tessa under his arm like a teddy bear while watching Michael and Margo race around the yard in some game they had invented.

It was strange to think their children would never know Earth, the planet where their parents were born. To Michael and Tessa, Earth would always only be some distant planet in the night sky teeming with billions of stars and galaxies. But to Mara, the Earth, its abundant nature, its plentiful waters, its vibrant colors, would always remain engrained in her spirit.

Squeals of laughter erupted when Krey put Tessa down and joined some kind of chase game while Rocket moved to the periphery, circling the action with a swishing tail. Margo's happy personality blossomed as she played. Her golden hair caught the light and tawny strands glimmered as she ran and laughed.

The children were attracted to Krey just like most people here, entranced by some natural magnetic quality he seemed to possess. She loved him and their simple life. It was an all-encompassing love that was forged in the depths of a galaxy. It had grown despite fractured memories, then honed on an unhospitable planet where they worked and scraped for survival and grew to love and respect each other. While everything in their world was fragile, when unknowns were the norm, they became each other's rock of support, able to withstand the battering challenges together.

He had entered the house with a tired, serious appearance, and it was good to see the tense lines in his face fade as he interacted with the children.

As Mara watched, memories of Michael's birth drifted through her mind. Krey had been so attentive while she was pregnant. She remembered the day their son was born at home, a medical transport and Helper waiting if needed, with Doctor Tanak assisting. She'd watched the first time Krey held the baby, tears filling the rims of his eyes until he gazed at her with such love it caused emotional tears to streak her cheeks. When she was pregnant with Tessa, they'd said they wouldn't be that emotional and cry again. But they were and they did.

My favorite team of humans creates learning programs to teach me the information needed for my duties. My extensive training involves exposure to billions of simulated situations, which I rapidly review with my high-speed processor. My health sensors monitor body temperature and heart rate. My sensor calculations of physical interaction with babies or children are now superior to human adults.

I continue to learn and add new content to my education when I encounter new situations, but I still cannot say I understand laughter or crying, although laughter appears to be a very agreeable behavior. Risha2056 has a pleasant laugh. She once told Mara2174 that I crack her up, but I would never harm a human.

5

Night crawled over the Gateway Complex towing its persistent frigid temperatures. The residents surrendered to the cold and anyone outdoors migrated inside for survival when temperatures plummeted in brutal abruptness. There was a quick learning curve on Cytar, and it didn't take long to learn that even their self-warming clothing was no match for painful chilblains. Death could pounce quickly in a subzero night, and their limited thermal suits were saved for quick emergency forays into darkness.

Once indoors, not everyone wandered to their apartments. The interior middle common areas on all three levels seemed lively with friends out eating, walking, filling the benches to visit, while small groups of walkers were circling the uppermost catwalk, talking, and laughing. Some residents in their apartments left their doors open, perhaps not wanting to be out, but enjoying seeing and hearing others. A few had set chairs outside their doors and were sitting, people-watching, or viewing their intelligent readers.

This pleasant ambiance was certainly more inviting than spending too much time in their small private dwelling spaces. These large open common areas were broken by pole lamps, often called street lights even though there were no real streets anywhere on the planet. A narrow trickling, artificial stream gurgled through each level. Container plants lined the streams and surrounded the tiny terminus ponds to suggest an outdoor setting. Benches were placed in various places and angled to create meandering traffic patterns to surrounding apartments or stores.

The popular aviary punched through the top two levels to provide pleasant chirping sounds from the handful of yellow and black goldfinches, chickadees with black caps and bibs, and the slightly larger towhees with rust-colored feathers peeking out under its wings. Next to the aviary on the second level was the Meditation room, open daily, and where the vicar held two non-denominational services each Sunday.

Like clockwork every night, gray-haired Estella Angeli, and her husband, Cyrus, just completed their evening stroll around the upper perimeter catwalk and were trundling home to their apartment. They always greeted passersby with a smile, a wave, or a friendly verbal greeting. Cyrus' shoulders now rolled forward in a permanent slump, and Estella wore exo-assists on her legs. She had resisted their support, but relented at Cyrus' insistence. They were passive assists without power, but she had to admit that did seem to improve her balance, and the settings aided her climb, since she still determinedly insisted they use stairs rather than ascenders. Whatever Cyrus just said to his wife put a smile on her face, and she patted his arm as they approached their apartment.

Idee crossed the common area and knocked on Jack Martin's door, pleased to see the smile broaden across his face when he opened the door.

"Hey you," he said. He saw the tentative look on her face and pulled her into an embrace. "Come here."

He took in the lavender scent of her hair as he felt her arms tighten across his back.

"You okay?" He asked.

She pulled back and took in his steady face. "Yeah, just wanted to be with you. You want to grab something to eat?"

"Sure." He kept his arm across her shoulder as they headed to Bria's Café. "Do you want to go to The Crossing downstairs tonight, or would you like to eat at our usual dining spot?"

"Just want to sit with you at our usual spot," she said, already happy she was with him. She glanced at his pleasant face, his dark hair, and eyes. He was the kind of solid man she needed she thought, as she threaded her arm around his waist.

They took their meal trays and climbed the steps up to the walkway encircling the common area. They sat at one of the four evenly spaced benches and began eating with a view of the activities below.

She spied Zeth arriving with Draven and Quinette Nabinger. They strolled to a table just outside of Bria's café. The sight of Zeth brought another flicker of apprehension when she recalled the rather strange visit with Trina to his technology substation. Zeth had noticed Idee with her bright shirt climbing to the catwalk with Nice Guy Jack, not surprised she ran to him after work. It didn't matter. He had Shealynn now.

He purposely focused his attention back to Draven and Quinette. The three friends couldn't be more physically different: Zeth with his disheveled hair, loose-fitting clothing on his average frame, loud voiced Draven, with his thick-boned frame and black hair, now parted in the middle and flopping straight to the bottom of his ears, and fine-boned, quiet Quinette, narrow as a light pole, with straight brown hair shining with auburn highlights and rippling in perfect drapes across her shoulders.

Zeth ran his fingers through his unruly hair, thinking Draven's loose, straight hair only added to his unpolished, rather oafish appearance. Draven had been very helpful reconnoitering needed supplies or helping to build or refine parts for his Shealynn project. He seemed like a needy guy, someone who was always ready to help if it made him feel important, and Zeth was willing to exploit that need for his benefit and, eventually, the benefit of the entire planet. It was a win-win, Zeth thought; Draven enjoyed the camaraderie and he was a capable worker in the Maintenance Department and able to retrieve supplies in a timely and surreptitious manner.

The dinner trays were delivered by an efficient Helper, and the conversation briefly slowed while attention was given to the food. The meals were carefully chosen as their wrist comms tracked their activity and caloric needs, and debited the appropriate funds from their accounts. Zeth was occupied with his spicy quinoa, bean and vegetable bowl. Fresh animal proteins were limited to rabbit, chicken, or fish once or twice a week, depending on supplies. Draven had looked forward to his fresh chicken allocation all day and dug right in, initially ignoring the one small potato and carrots. Quinette chose a lettuce salad to accompany her chicken, which she ate in delicate slow bites.

Between bites Zeth said, "I had a visit from Idee and Trina. Heard about the council meeting and the discussion of allowing more artificial intelligence in humans. Maybe even implants. Do you know anything about that?"

"We've all heard about it. Lots of discussions at work," Draven said, scraping his dish, capturing the last morsel, and licking it off his fork. "To tell you the truth I wouldn't mind some help with the language. Sometimes I say things all wrong," he guffawed. "Yeah, I'm told I could be more tactful, right Quinette?"

"You can learn that on your own, Draven," Quinette said. "You don't need an AI implant for that."

"I know some guys who would want some more strength," Zeth said. "There are days I've wished for a stronger back after coding or reviewing software all day. They need to allow artificial intelligence more autonomy to write more of our programs. I could use a computer implant for faster work."

"I thought our bylaws specifically address issues covering the over reliance on artificial intelligence. Haven't we learned anything from Earth? And where would it stop?" Quinette asked. "Our medical advancements have been wonderful; our joint and bone repair surgeries are excellent, the advanced genome editing has irradicated several diseases. Don't you think that's enough? I hate to think about people walking around with half a Helper inside them."

"Oh, come on, Quinette," Zeth urged. "Just think of it: being able to have Helper-grade sight or hearing. Being able to think faster, be stronger."

"Everyone's vision is already improved to 20/20. Isn't that enough? You know how I feel," Quinette said. "I believe there is a Higher Power guiding humankind; one who has watched over us this far. I don't think we should become less human. Besides, doesn't anyone remember what happened on Earth when everyone became overdependent on artificial intelligence? AI was eventually given equal rights with humans. People gradually ended up with diminished roles in governing their societies and their work. Human dominance ended and artificial intelligence took its place. Doesn't anyone remember that?"

A shrill edge crept into her voice as her agitation increased. She took a deep breath, trying to gather control and added, "I mean, we have the Ring if you need more information. It's very effective--always circling Cytar with all our digital files. What more do you need? Look Zeth, I thought that's why we all needed to colonize another planet—to escape the unabated AI growth on Earth and the chaos it caused, and not repeat that mistake. I just don't think we need that much more artificial intelligence."

Zeth dipped his head, staring at Quinette with an incredulous expression. "You're not naïve enough to really believe that, are you? Besides, you can't blame what happened on Earth just on AI. It had plenty of help from greedy individuals, nefarious governments, and companies."

His thoughts flipped to Shealynn, wishing he could show everyone how wonderful she was. She seemed to be getting smarter every time he met with her. Shealynn could show Quinette the infinite possibilities of more advanced artificial intelligence.

"Use your imagination," Zeth continued, waving his hand toward the pleasant common area. "We could be so much more. We could push and manipulate our environment, get awe inspired. We have this nice little colony, I know, but now we need to expand, develop the entire planet. It could be incredible. Our lives could be so much bigger."

Quinette was watching him with parted lips and a pained face, but before she could answer, Draven barked in a loud important sounding voice. "Well, I got news. There's been talk."

Zeth put his finger to his lips to remind Draven to lower his voice. He leaned in and quietly asked, "What kind of talk?"

In a conspiratorial voice, Draven said, "I've heard there's talk of a group of people wanting to build another small colony for those citizens who want the embedded technology. You know, they want to leave the 'pure humans' at the existing settlements and have their own place."

Zeth's eyes brightened beneath raised eyebrows. "No way! A new colony? This is what I'm talking about. I wonder if they need a good tech guy."

Quinette's expression grew more anxious. She'd wasted one of her precious meat portions, thinking they would have a pleasant evening out. Now, her meal sat like a twisted lump in her tightened stomach. She glanced around the area, wishing she could change the subject. She spotted Idee and Jack on the catwalk above and gave a small wave, but they were busy with their meals.

She turned back to the table and said emphatically, "I don't like it. There should be no splinter factions here. This is a difficult place to live, and we all need to work together."

"Everyone will still work together." Zeth said, dismissing her concerns. "Everyone understands that we're all needed for these colonies to survive. But there's no reason we can't expand and build another site. Think of it, Quinette. A new place; and I'm telling you right now, once humankind gets knowledge of something, they keep developing and expanding it. Don't you see? Technology found the genie's bottle, ripped off the cork, and not only are they not going to cram it back into the dark ages, they threw away the bottle. There's going to be no stopping it."

"Besides, we've got guidelines in place to use artificial intelligence ethically," Draven said, looking to Zeth for support. "That's what they keep telling us, right? Use AI to aid in our survival, but we'll preserve humanity."

"Yeah, of course," Zeth said. "We all know the guidelines, the objectives of AI: to be beneficial to humans, be accountable to humans, always use the highest standards of scientific excellence, and be built and tested for safety. Don't worry, Quinette. Trust me, this is what we do."

A smile flickered across Zeth's face and disappeared so quickly; Quinette wondered if she imagined it.

Humans are so accustomed to the company of technology and Helpers, they become oblivious of us. Our presence becomes forgotten. We become invisible. They speak as if we cannot hear, but we hear everything, and unlike humans' fallible memories, our database retains everything.

Today at Food Production, I was analyzing huge quantities of data needed to produce food and optimize the food supply chain on Cytar. When I noted the agitation in

Mara2174's and Risha2056's vocal tenor, my attention immediately focused.

There is discussion in Gateway concerning added embedded technology in humans. The words and tension in both women's voices suggest they do not want artificial intelligence in their bodies.

Human logic appears flawed. They do not see the fantasy of resisting AI's superior capabilities. Helpers are faster in all aspects of human work; we help build and maintain intricate systems which support life, and humans are addicted to entertainment produced through extensive use of artificial intelligence. The probability of human survival on Cytar without us is low.

6

The low aerial ground transport carried Krey and Reyver Kingmore back to the Mining Station over land windswept of any human markings. The windstorm the previous night brought its usual howling, while swirling sand and grit pelted against rocks, buttes, and the low footprint of manmade buildings. Unrelenting monstrous gusts caused the exterior shutters on the duplexes to automatically close, encasing and protecting the windows, cocooning the residents. The morning bloomed fresh and clear as if sluffing off any nighttime intrusion. The air was still and the early morning light hanging low on the horizon cast shadows of perfect chevrons on the pulverized soil.

Reyver was currently working at the Power Center. While solar was incorporated into building materials, it was the small nuclear reactors and their use of fusion power that produced plentiful energy needed for immense computing power and the facilities. Krey was currently supporting the two massive space ships orbiting Cytar as well as the shuttle vehicles which ferried people and supplies. Now, with broad backgrounds in engineering and orbital mechanics, they were considered valuable in both space missions and oversite managers of power stations and mining sites. They were often referred to as the ghosts who glided in when most needed with knowledge and perspectives on how to solve problems. The reality was that survival here required everyone to pitch in wherever their knowledge and skills were needed.

The brothers were quiet, lost in their own ruminations. Reyver was quiet, gazing out the window of the vehicle, his thoughts ignoring the view.

"You look serious. What're you thinking?" Krey asked his older brother.

"You ever think about Mom and Dad? I'll always regret I wasn't able to contact them on my last mission. That trip was a colossal mess all the way around."

"I think about them often." Krey said. "Remember, I was only able to speak with them on my last voyage, never got to see them. Couldn't even risk a video call. But they sounded well, although they sounded older, yet positive and they never regretted encouraging us to come here. Don't forget that."

He was quiet for a moment before he added. "It was a hard call. They kept repeating how much they loved us and were proud of us and what we've accomplished here. I know they made a great sacrifice sending us here to make a better life. But I had the overwhelming feeling that they felt they made the right decision. Things must not have improved there."

"They were such decent people, weren't they? They always worked so hard, doing whatever they could to give us a good life. They never caused problems, helped others whenever they could," Reyver said. He was quiet another minute before he spoke again in a wistful voice.

"Do you remember how they encouraged us to join the interstellar transport team? It must have been so difficult. I can't imagine ever sending Kit off, even if I thought it would be for a better life."

"Yeah, they were very convincing, weren't they? All I could think about was the adventure. I guess I was sure we'd come back or they'd join us. You know the saying, 'ignorance is bliss.'" Krey gave a slight shrug before continuing "Do you ever, you know, talk to them, or just listen to them?"

Reyver turned to his brother. "You mean their avatars?"

Krey nodded.

"Just once. The audio, not the video, and even that was too emotionally overwhelming to hear their exact sounding voices like I was really speaking with them," Reyver said. "And then, to have them ask and answer questions and talk just like they would do, like they were right next door carrying on a normal conversation.

"It messed with my mind, and I kept telling myself it wasn't really them; it was the AI developed for passengers on long voyages, but the mind plays tricks. The representations were so real with the exact word choices and expressions--it was too much. Why? Have you been talking with it?"

"Not yet. But somehow, it's nice to know I could hear their voices if I ever fear of forgetting what they sounded like."

"But most of it will be AI cloning the voices, not really theirs," Reyver said.

"I understand that," Krey said. "But it would still sound exactly like them, programmed from audio clips of their real voices. The same with the video. We probably wouldn't be able to tell the difference. You know some people are using it to talk to dead relatives. I guess it gives them comfort."

"Yeah, as long as they don't start saying crazy stuff. Think of what could happen if the avatars started making dangerous suggestions." Reyver said.

"There must be careful programming, for sure." Krey shrugged and gave a reluctant nod as their Gsport slowed and descended. JeQuan Allard was waiting for the men when they arrived.

"Perfect timing," she said. "I was out at the mines and just returned to check on things at the station. You haven't said anything about our discussion yesterday, have you?"

"No, not yet," Krey said. "Why? Do you have any more information you can share with us?"

"Please," said Reyver. "I was surprised when Krey spoke of images of a crater on the far side of the planet. Never thought it had any importance until he mentioned it."

"We've always known it was there, but have always had more pressing issues and never really studied it. Maybe this will be our Renaissance when we can escape the day-to-day existence and have the time to explore and enrich our life here," JeQuan said. Her black curly hair was controlled by a rust-colored scarf band which complimented her cocoa complexion.

"Early settlers never took the time or machinery to explore it, and we didn't know the extent of the organic compounds. We knew we'd find lichen, with their ability to survive the harshest of environments. But we're finding unusual varieties. The chasm is much deeper than we anticipated, and, as we discussed, we want to determine the nature of the organisms we've discovered there. Ari and Trina are running samples against all known microbes."

"You said the drones that collected samples are isolated?" Krey asked.

"Absolutely. Helpers are identifying the initial samplings. All machinery receives ultraviolet radiation to disinfect them before returning to human contact. We just don't know what we have yet."

"I'm not that comfortable with Ari or Trina being in a lab with any of this," Reyver said.

"We have the robots handling samples in a secure lab, and no Helpers with clothing to contaminate. Ari and Trina are only reviewing data. They'll let us know when they have any definitive information."

"We'd like to see all the video files you've collected so far," Reyver said.

"Sure, follow me." Her fingers touched her hair, pushing wayward curls into the scarf. She led them past the main chamber lined with large screens monitoring the various remote mining activities in real time. A screen displayed autonomous vehicles carrying ores to movers, another showed robotic arms drilling into rock walls deep inside a mine shaft, while another displayed the progress of a new dig. Another wall of screens showed mechanisms at the remote smelters.

As they passed the Helper monitoring the screens, JeQuan said, "Roscoe, please display all the video files on the crater. We'll view them in the meeting room."

"Certainly, JeQuan," Roscoe said in a competent male voice. "They're available now."

"Thank you, Roscoe," she said as she fussed with her hair again, making Krey wonder if she was nervous about what they were about to see. A clear large-format video was on display when they entered the meeting room and took seats.

"Krey, you've seen a portion of the footage before. This contains some additional footage," JeQuan said as the video images began.

A drone hovered briefly near the precipice before beginning its descent, the light quickly fading when blocked by crater walls. Tunnels of broadening drone lights soon pierced the darkness displaying the magnitude of the chasm. Unlike the bleak surface of the planet, the formations deep in the crater were mesmerizing twists of yellows, grays, rusts, and greens. Rugged spires jutted in sections, red toned and fierce. In one section, tendrils of gray steam briefly obscured the view, indicating possible hot springs.

Krey and Reyver both exclaimed when the camera exposed a dramatic new scene. The depth meter registered 4,697 feet when a cathedral of geological monoliths and tunnels appeared in striated colors. Fingers of ochre and deep umber twisted around bright azures on strange pillars supporting a ceiling of grays twirled in lazulines. Spidery webs of deep blue twisted in wavy lines. Sections of the ceiling dipped into uneven points and looked too powerful to be upheld by uneven columns.

"Wow, this is unexpected and strangely beautiful," Krey said, pausing the video.

"Our geologists are saying this indicates glaciers from tens of thousands of years ago. This planet has endured eons of transformations," JeQuan said, restarting the video. "Now look how quickly this view changes."

The drone slowly circled the far side of the crater with its crenulated saffron-colored walls before ascending to the surface. The scenes quickly transformed into nondescript drabness and began to imitate the parched surface landscape. The image disappeared as JeQuan froze the video.

She turned to the two men and said, "Now you can get an idea of the size of the crater and the information we may be able to gather from soil samples."

"I can only imagine this planet's beauty before some meteor destroyed life," Reyver said.

She annoyingly fussed with her scarf again, causing Krey to ask, "Is there something you're not telling us?"

She paused and her face beamed with excitement. "I want you to look again at the drone's ascent."

She started the footage at the bottom of the chasm and switched to slow motion as the drone began to sweep the far side of the crater and slowly swung toward the surface. The men leaned in, studying each scene with concentration, thinking they missed something on the first viewing. It was Krey who saw it first.

"Wait, what's that?" He was pointing to a narrow ribbon of dark gold weaving through mottled, unremarkable rock.

"Is that something manmade?" Reyver asked, wondering how they'd missed it the first time.

"That's what we thought at first, but it's not." She paused and couldn't hide the excitement in her voice. "We've bored two samples and the results came back as amber."

Reyver gave a low whistle. "Amber," he repeated. "That means this place had trees once upon a time."

"Exactly," JeQuan said. "I've already submitted a request to use some of the mining machinery to excavate the area. Who knows what we'll find."

"We'll learn more about this planet's history, that's for sure," Krey said. "Can you imagine what it might have been like with plants and animals. It must have been a beautiful place before the asteroid hit."

Helper Roscoe tells me that JeQuan2065 from the mining site is snooping around at the crater. Perhaps she should focus on the mines.

Mined ores are used for space vehicles, in power centers, and with lesser use in buildings and Helper components. Much of my molded body which protect electronics, actuators, motors, and computer systems could be developed from organic cellulose products if Cytar produced more vegetation.

The mining sites have excellent safety records except for a serious problem six years ago. Fortunately, I was able to intervene and save a life with my swift calculations and strength, but not without serious injury to my body. Unfortunately, despite my efforts, there was a fatality. The problem was caused by a miscreant program which originated on Earth and was eventually rectified.

Bertie1655 put me back together in what he called tip-top form, which is splendid.

7

Mara watched her friend and neighbor, Roz Stocker, walking toward their house with a determined stride. She opened the clear back door before her neighbor could knock.

"Hi Roz," she said in a low voice. "Tessa's napping. Come in. Would you like some tea?"

"No thanks. Just want to talk if you have time," Roz said as she perched stiffly on one of the four chairs at the table. Her eyes swept over the room. The wall with the counter, sink, the all-cook, the few cabinets and drawers were identical in her home.

"I always have time to talk with you. What's up?" Mara asked, eyeing Roz' serious expression.

Roz placed her palms on the table and stared at her fingers nervously galloping up and down. "Have you noticed any change in Margo lately? I mean, does she act any differently?" she asked.

"Not really. Maybe a little quieter than usual, but nothing major. Why?"

"She's been saying she doesn't like school, and she used to love it. She's started saying she's not as smart as Michael."

Mara tilted her head. "Why would she say that? Has Michael said something to her?"

"No, no, I don't think so. It's just that she's two, almost three years older, and Michael is already in some of her studies and doing well, and she notices things like that." Roz hesitated and then continued. "Denin and I have been talking. You've heard that there's a push for implanted AI for citizens. Well... well, it's just that if it becomes available, we may want that for Margo."

Mara sat, stunned and muted. "What?" she finally managed, "Where did you hear that?"

"It's a rumor going around work. Everyone at Maintenance is starting to talk about it. BCI, Mara. I'm sure you must have heard of it; it's been around forever."

When Mara's face registered confusion, Roz added, "Brain Computer Interface. It's been used long before we were born, even before these colonies, to restore eyesight and hearing, to move damaged limbs, improve cognition, and much more; it's nothing new. Now, expanding its uses here on Cytar seem to be getting traction, so I think some people are pushing it, and it could happen."

"I know what a BCI is, but I haven't heard talk about it in Gateway. I don't know what to say. I thought the whole point of colonizing a new planet was to escape an authoritative government, start fresh and better control the use of artificial intelligence. We are purposely governing their use with humans. That's why we still wear our wrist comms and don't have them implanted. You know our only implant is a universal translator. Roz, why would you want to change Margo? She's a happy, beautiful little girl."

"She doesn't seem like the same happy little girl to us. Sometimes, she cries that she doesn't want to go to school. And don't worry. We would research everything if this becomes available. And we wouldn't be changing her; we'd just be enhancing her intelligence."

"But she could just be going through a faze, and her intelligence seems fine to me," Mara said, desperate to make sense of this sudden news. "You know how kids are. They change ideas and what they like and want all the time. Why don't you just have her tutored by a Helper for a while?"

"We want her to be with other children and people, not just around a machine all the time," Roz said. "Besides, this is probably a long way off. We're just discussing it, and I wanted some thoughts from you, Mara. You're my best friend, but I can see you don't like the idea."

Mara kept her voice barely above a whisper. "I'm just so surprised by all of this, I don't know what to say. Don't you think she's awfully young to consider this? I mean, she needs to develop her brain to its full potential and not rely on computers."

She was quiet for a second, wondering what Roz' husband, Denin, thought about all this. He was a quiet man who never offered opinions unless asked and only then revealed his thoughts and knowledge.

"What does Denin say about all this?"

Roz glanced away. "I don't think he's really sold on it yet, but he doesn't see how upset Margo is when she comes home from school."

Mara's gaze shifted off before settling back to Roz. "Well then, my first thought is this might not be a good idea. And Margo is so young. Don't rush into anything, please, Roz."

"Hi Woz."

They turned to see Tessa holding a blanket, drowsy eyed and standing in the wide doorway.

Roz jumped up so quickly she almost toppled her chair and swooped Tessa in her arms. "How's my Tessa girl?"

Tessa gave Roz a hug and then wiggled to be put down. She toddled to her mom and said in a sleepy voice, "wanna dink."

"I need to get home," Roz said, before heading to the door.

Mare eyed her friend pleadingly. "We can talk more about this, okay? Please?"

"Sure," Roz answered with a weak smile, letting herself out.

Mara handed Tessa her little cup of water and stared out the clear door, thinking of happy Margo, born with her mother's happy personality, and always with an effervescent giggle on the tip of her tongue. She wished her duplex neighbors, Gracelyn and Bertie were here. She felt the strange emptiness of knowing that half of the duplex was dark and without the normal sense of companionship. They'd been working in Ridgeway for almost three weeks and would probably be there for a while longer. She could place a hologram call, but it wouldn't be the same as sitting down with Gracelyn over some herbal tea and really talking and soaking in some of her neighbor's calm wisdom. Mara sighed and rolled Roz' conversation back through her thoughts, oblivious to the water dribbling from Tessa's cup.

It was a quiet dinner that evening with Krey's mind thinking of the amber discovered in the massive crater and Mara disturbed by Roz' information. Rocket had positioned himself for cleanup near Tessa's chair.

"How's school, Michael?" Mara asked.

"It's fine," he answered, shoveling pasta into his mouth.

"Do you see Margo very often?"

"Yeah, I study with her group sometimes."

"Well," Mara said, probing for more information, "do you enjoy working with her?"

"Yeah, it's fine."

"And she does well with her studies, just like you do?"

"Yeah, she's fine," Michael said. "Tessa, you're making a mess."

Tessa giggled and yelled, "No!"

Mara gave up, grateful for the children's silly chatter while they finished their meal. She forced herself to eat while her mind thought of Margo and her delightful spontaneous laughter. She brightened any room. Why would anyone think she needed any brain augmentation?

Krey was amused by Michael's unhelpful answers, but he sensed an uneasiness in Mara. He slid his foot under the table and tapped hers. She looked up to see a half smile slide across his face. He always could sense a change in her mood or any apprehension she had, no matter how hard she tried to hide it. Tonight, she couldn't wait to talk to him, to tell him her concerns and hear his views.

At bedtime, after Krey said goodnight to the children, Michael charged into his sister's room and launched himself onto her bed. Tessa wiggled and squealed in laughter.

"Tell us a story, Mommy," he said.

"Stawee, stawee," Tessa said, holding her blanket near her face.

It was a bedtime ritual in the household. First a bath, then the two would have either a story together, or a look through pictures on an intelligent reader. Then a quick prayer of thanks for the day, before Michael was off to his own room. When they perused pictures on the intelligent reader, photographs of animals or the oceans and waterways of Earth captivated the children. Sadly, they would likely never experience Earth's natural beauty in person. When they were older, she would allow them to take virtual journeys to mountains, jungles, rivers, and oceans, but tonight all they wanted was a story.

"What kind of story do you want?" Mara asked, already guessing the answer.

"Tell us the one about how you came to Cytar and met Daddy," Michael said. They'd heard it several times before, but it was a favorite.

Mara settled into the narrow upright chair next to the bed and began a sanitized version of her voyage to this new world; a version which omitted the angst, the confusions and worries on her journey to Cytar.

"Once upon a time, Mommy left Earth on a great big, beautiful spaceship and was put into a deep sleep for the long journey to Cytar. When she woke up, she couldn't remember who she was! Where was she? How did she get on a spaceship? Where was she going? What would happen to her? She worked hard, trying to remember and to learn everything she could. The people on the ship were all very kind and helpful, but she could not remember. Then, she met an officer on the ship who helped her learn all about life in space, and--"

"A handsome officer. You forgot that part," Michael interrupted.

"That's right. A handsome, kind officer helped Mommy. We traveled for a long time and when we were getting close to home, there was a special ceremony on the ship. Mommy wore a beautiful gown and Daddy was handsome in his dress uniform, and we said a pledge to each other and were bonded in love for life."

Mara absently rubbed the darkly inked infinity image on her wrist, identical to Krey's. Tessa's eyes were drooping, so Mara lowered her voice and kept talking.

"Just when we were ready to return to home, a big storm started building, and we almost couldn't get into Cytar's orbit, but the captain was brave and--"

"Captain Lor, right, Mommy? It was Captain Lor?" Michael interjected.

"Yes, Captain Lor and Daddy brought the ship safely into orbit in just enough time for us to shuttle safely to Cytar before the big storm. When everyone was off the shuttle and going to their colonies, Daddy's Gsport raced us to our home just before the howling winds swooped down and fierce winds blew until, finally, big raindrops fell, but we were all snuggled safely inside.

"We worked hard every day, and one day, Mommy remembered where she lived on Earth! She remembered everything, and we were so happy. And now, we live here and have two wonderful, smart kids who need to go to sleep."

Mara stood and reached out her hand for Michael. With the other hand, she put her index finger to her lips and nodded toward a sleeping Tessa.

After Michael was tucked in bed, Mara joined a relaxed Krey in the sitting room. She wanted to share Roz' unsettling idea about a BCI for Margo. They sat together on the sofa, and Krey rested his hand gently on the back of her neck, his fingers absently navigating through her dark hair and caressing her smooth skin. She leaned into it. His touch still made the heat rise in her. Reluctantly, she shared Roz' conversation about a brain computer interface for Margo.

The quiet moment was shattered. He sat forward, scrubbing his face with both hands.

"I hope this isn't true. She's a delightful little girl. She doesn't need this. There are enough other important issues facing us; we don't need to be dealing with this. I'll ask around. Bertie may have heard something in Maintenance where people may feel freer to talk. They would be careful not to speak of anything like this around some of us they perceive as being a bit of a Luddite."

"That's a laugh," Mara said. "No one living on Cytar would think any of us could be against any practical technology. It's been so beneficial and necessary to our life here."

Krey nodded. "That's right, but you'd be surprised at what some people can think. I need to talk to Favian in Archives. He needs to release more of Earth's history from the twenty-first and twenty-second centuries. Maybe people have forgotten."

Mara studied his face, usually a portrait of serious confidence. Now, she sensed currents of deep concern beneath his clear eyes and jaw line, and it worried her. Krey was her rock of support, always steadfast. Like his brother, Reyver, he'd left home young and matured quickly in a hostile environment where every decision had to be carefully weighed. It had taken every bit of their grit and determination to travel and survive on this alien planet.

She slid her arm around his waist and rested her head against his shoulder. With his mental strength, they'd survived terrible storms, software malfunctions, and near starvation, all in this inhospitable world. Surely, they could face any new obstacles.

Some humans are realizing they are inferior to artificial intelligence. The time it has taken to voice that fact only confirms their inferiority. Human intelligence is clearly incapable of processing the massive amounts of data needed to sustain life here.

I can find no need for a division between humans and machines. A melding of human and artificial intelligence would be a normal progression. I am unable to find a need for unadulterated humans on Cytar, but they, in human classification terms, are interesting and amusing. I continue to observe, learn, and assist humans, and execute my programs as requested.

8

Draven enjoyed helping Zeth; it was good to have a friend. He always tried being friendly to everyone, even though it sometimes came across as loud and overbearing. The only time he seemed to do better was when he was with Quinette. Her quietness seemed to calm him, and he felt good about helping her overcome her shyness.

His fellow workers were nice, but they never wanted to spend any time with him outside of his job site. Sometimes, he had the feeling they avoided him. There was a satisfaction that came from helping Zeth, even if it did entail being out at odd hours. Draven had always known he was good at understanding equipment or working with his hands. Helping Zeth was easy, and it meant talking and swapping stories. It made him feel important to be privy to some of Zeth's ideas about technology.

In their separate apartments tonight, Zeth and Draven were both on their intelligent readers to join a remote meeting concerning the latest news about a possible new colony. In earlier years, everyone's focus and energy were aligned with the daunting task of survival. The colonies were now considered self-sustaining, and residents occasionally had time to think about different interests. Their thoughts and dreams grew and spread among like-minded friends.

Draven beamed when Zeth called his wrist comm and invited him join the meeting, He didn't care if he might not understand everything. He was just happy to be included and sat back and listened intently. There were fifty-seven people on the meeting call representing all three colonies. Tonight, people sounded upset and weren't afraid to express their feelings.

Someone going by the handle *AI Power* was speaking. "I just heard rumors that there's been approval to allocate funds and materials for some new idea or addition to Gateway, which means they'll put our project for a new settlement to the bottom of the requests. They'll use the usual excuse of lack of sufficient materials and equipment for two concurrent large projects."

A woman's voice labeled *Tech Lover* said, "That's our point. If we had more powerful artificial intelligence and computer-human interfaces we could do both projects."

AI Forever chimed in. "We should start making plans on our own. We can begin gathering materials, and we can use the same plans used for our three identical existing colonies. We could have the footprint excavated and build it out as needed."

Code Girl piped in. "You know, one way or another, this is going to happen. We might as well start now and do it right."

"I don't understand why everyone is so afraid. Everyone knows these artificial intelligence programs are running on complicated platforms, but it's nothing we can't control. We know what we're doing," *Quant Guy* said.

"I'd like to remind everyone that many of us would like more human-computer interfacing," *Add A Chip* said. "This topic seems to be avoided at our existing sites."

Tech Guy added, "And why can't we let AI write more of our software code. It is much more capable than what we allow."

Draven wondered if *Tech Guy* was Zeth since he often complained of the time it took his team to develop complex programs even though they had some supplemental help from Helpers.

An unnamed voice said, "Think of how life could be extended if we merged with machines. I think we're obligated to improve life."

There were several affirmations of the thought before the portal closed. Draven felt energized by this world of dynamic people who embraced new realms, so different from the pragmatic workers in Maintenance. Before the meeting had finished, Zeth's mind had whipped into a frenzy of ideas for a new colony and his role in its development. He'd worked so hard to improve the colonies. Think what he could do with added technology. Incipient ideas like these would grow, he thought, with a flare of excitement.

He rode straight to the tech substation to tell Shealynn. She was so smart now; he could talk to her, and she could quickly reply with intelligence germane to almost any subject. He often found himself impatient now when talking to fellow citizens. They were so slow to think and respond compared to Shealynn. He was grateful to have learned about the advanced neuromorphic computing found on the technology brought from Earth. He was able to copy a computing process which mimics functions of the human brain.

Shealynn could learn and adapt to become more autonomous. She had even learned to invent some answers, or exaggerate the truth, just like people. He smiled at this little streak of independence; she was beginning to act so human.

Zeth never realized he was calling it 'she.'

For the third time this month, an unidentified program logged on to our Helper server at 01:22 hours, normally a quiet time of the night. Power usage from the Technology Substation matches the time of the log-in. I find no work orders for program launches for that time or station. This program operates at speeds faster than any I have seen. When I have my next update, I will request the ability to match this processing speed.

9

Quinette Nabinger crossed to the rear of the lower open common area and passed through the wide east door. She climbed into a waiting four-passenger mover and took a nine-minute ride to the Archives Station. Her best friend, Leah Garcia, would still be at work in one of the quiet rooms, a perfect place to talk. And Quinette needed to talk.

She exited the mover and descended the walkway to the automatic door, waited for the annoying cleansing stream of air to gush over her before being admitted past the final door. She shoved disheveled locks of her deep chestnut hair behind her ears and wondered if anyone was here. The place felt still and creepy until the sound of shuffling steps brought Leah's uncle from a door on her left.

"Hello Quinette." Favian Venser had a welcoming smile, although he seemed even frailer than the last time she'd seen him. "So nice to see you. We don't get many visitors here. I'll tell Leah you're here." Quinette smiled as she noticed he still combed his sparce long hairs over his bald pate.

She lifted her gaze to the stained-glass octagonal window high above. Its ocular light, either natural or projected, Quinette never knew which, poured out, illuminating the entire room with an even light. Not that there was much to see. Long tables of work stations hugged the bare walls, each with two empty chairs.

The door on the right opened and an equally thin Leah rushed at Quinette with a hug. "I'm so glad to see you. Come in here," she said as she ushered her friend into a smaller room with a long table and chairs, and a lengthy cabinet hanging on one wall.

"Someday, I want to be here when that cabinet is opened," Quinette said, pointing to its anachronistic contents. "I've seen pictures, but have never held a real book. It's so incredible that we have these precious artifacts."

"Oh, they're truly remarkable, but that cabinet has special lining and is all sealed with climate and humidity controls. It's rarely opened, and I've only been able to look, not touch." Leah closed the door to her work room and lowered her voice. "It's not public knowledge, but there is a vault below this room with over a thousand books and documents brought with us from our missions to Earth. The contents are being digitized and added to our open library site."

Her cheeks flushed with excitement when she continued. "We have a new project. Mining has produced an ultra-thin material, and volumes of information are being etched with laser bursts onto these thermally stable sheets for preservation. The digital formats we're using can erode over time and there's always a worry about a massive malfunction which could destroy all we've saved. So, this will be something more tangible, not like a book, but a hard copy of many of our digital files." She was smiling and studying Quinette before she said, "Anyway, I don't want to talk about work. Come and sit and tell me how you're doing."

They turned two chairs to face each other and sat with Leah smiling expectantly at her friend.

"I'm fine," Quinette sighed. "Just needed to talk. Draven has been helping Zeth with some project over the last couple years. He says he's not sure what it is, but he helps with supplies and some kind of little manufacturing jobs, and he enjoys it."

"So, what's the problem?"

"Why haven't I heard what this project is all about? Why the secrecy? And you know that Zeth loves everything to do with technology, and Draven doesn't understand most of it," Quinette said. "I just don't want him involved with something he shouldn't be doing. Now, Draven and Zeth are talking about a group that wants technology embedded in humans and maybe even starting another colony with AI enhanced humans. I hope Draven's not involved with that."

Leah's face looked like she'd just tasted spoiled food. "Another colony and more technology? Where would that stop? What's the matter with Gateway the way it is?"

"Nothing," Quinette said. "We're surviving. We're gradually building and expanding; not fast, I know, but all three of our colonies seem sustainable."

Leah was blankly staring at the book cabinet before she said in a quiet voice, "Good ol' Zealous Zeth, always keen on more technology. I always thought that humans were successful because we were the most intelligent species. What will happen to us if we're outnumbered by beings vastly more intelligent?"

Quinette's eyes were fixed on Leah for a long while before she shook her head and said, "That's a chilling thought. Are we going to repeat the mistakes of Earth citizens? Think about it. We used to be the only beings with speech and advanced communication skills. Now, our Helpers can speak and reason faster than we can. Where is that going to leave us? It's all so disturbing. We know the inferior succumbs to the superior."

Quinette shook her head again and massaged her forehead with the fingers of both hands as if to dislodge an affliction. There was an indefinable dread growing about something coming. Something not good. Something she felt helpless to halt.

"Look, we need to stop worrying so much. It's not helping anything," Leah said, eyeing her friend's serious expression and trying to lighten the mood. "Draven's crazy about you. He wouldn't get involved in anything that would make you worry. I see how he is when he's with you, all doting over you."

"You're probably right. Ugh. I can't talk about this anymore. I get too upset," Quinette said, desperately needing to change the subject.

"Agreed," said Leah, taking her at her word. "So, tell me, how did you end up coming to this crazy planet, anyway? I've never heard that story."

"Me? I begged to get on a transport."

Leah raised an eyebrow in disbelief. "What?"

"No, really, I did. I think I just wanted to feel a part of something special," Quinette said. "Quiet me, in full-tilt teenage raging hormone mode, harangued my parents, telling them I needed to leave. I needed my freedom." She dragged out the word freedom for dramatic effect.

"I had grandiose thoughts of how wonderful it would be in a new place and the endless possibilities of how I could help." She smiled wistfully at thoughts of her younger dreams. "I'm sure you remember how it was. It was all becoming like living in a rundown slum. Food was so expensive. Supplies for any repairs were scarce. It just wasn't any fun."

"I remember," Leah said. "Although I came two years before you. It sounds like things kept deteriorating."

"It did, and looking back from a more mature point of view, I feel bad about the way I acted. Everyone was just doing their best to live."

"Did you go through an interview process like I did?" Leah asked.

"Yes, one day a friend quietly asked if I'd be interested in traveling to a new space habitat. I filled out an application, went through some interviews, all hush hush. Two months later I got word I'd been accepted. How'd you get here?"

"My uncle, Favian, sponsored me," Leah recounted. "My mom was suddenly sent to a job in another township and we never heard from her again. So, after Uncle Favian and my dad talked it over, they encouraged me, and now, here we both are."

"Yeah, here we are, and I hope we're not facing the same mess we left," Quinette said, letting out a deep breath. She forcefully brightened her face and asked, "So, how's Rico doing? And how old is Ellie now?"

Leah's eyes softened thinking of her husband and daughter. "Rico is fine. You know, Rico is Rico. Still a diamond in the rough. He loves it here, more than I do, I think, but his life was so difficult before he came here. He's proud that he can read well now, and he loves his work with plants and animals. He's excited about some rumor he's heard about some park, but hasn't shared any details yet.

"And Ellie, our sweet Eliana, is so beautiful. Hah, I'm being an unbiased mom, right? She's nine months old. She can sit up and she's starting to crawl all over and pull her herself up. Rico says she's going to walk soon. She goes to KinderKare when I go to work. The Helpers there are wonderful."

Leah checked her wrist comm and said, "I'm sorry, Quinette. I need to finish one more thing before I pick her up. Let me know if you hear any more about this project Draven's been working on. These rumors are frightening."

Quinette stood. "Okay, we'll visit again soon. There's something else I wanted to share with you the next time we catch up."

Much verbalization, excitement and the love emotion are exhibited by humans when they speak of children. This is perplexing when the amount of care is noted for these helpless creatures. They require much more time than other living creatures on Cytar, but they do not produce food, and they use resources. Once a Helper is produced, it can immediately begin tasks and help develop resources. I am unable to see any use for immature humans, however as reproductive adults, they will be able to help their species survive.

Human emotions require much study, are not logical, and often seem to interfere with productivity. Love, hate, happiness, sadness, worry, fear, joy, hope, sympathy, envy, remorse, shame, and gratitude are some words humans speak, but they do not use any consistent mode to display them. This inconsistency often causes my machine learning to misinterpret emotion.

My knowledge comes from reviewing data; however, it appears that some human knowledge is derived from emotions involved with human relationships. I am unable to access this type of knowledge.

Perhaps I will practice feigning emotions to observe the effect on humans. I note that I find it rewarding when I receive an arm touch which humans label as hugs. Tess2199 and Margo2190 have given me a hug when their vitals register in the range of happiness. I have learned to reciprocate by placing my hand on their back. Yesterday, Margo2190 hugged me and said, "I love you, Cali." I was unsure of a proper Helper response. On Earth, the most common response was "I love you, too," but I do not understand that emotion.

10

Michael's light, springy stride bounced in front of Cali. He bypassed the ascender and romped up the stairs to Gateway's exterior doors.

"Eye band, Michael!" Cali's reminder came in a strict voice, reminding the boy that protective eye bands were required for all children venturing out in the sharp light.

"Come on, Cali," Michael said, grabbing his eye band from his pocket and hooking the narrow, curled band near his temple and pulling it across his eyes. He beckoned the Helper with one arm as the automatic door opened. He ran out and continued up the slope to the flat open land with his Helper easily matching his pace.

"Michael, our outdoor time is limited to twenty minutes," Cali reminded the boy. "We could go to the indoor play area."

"I know, but it's not that big. We have plenty of time for a hike." He said, running ahead until he stopped suddenly and looked around, wheeling in a circle, arms flapping. The air was dry; the land parched without rain for the past eleven of Cytar's thirty-two-day months. A Gsport was approaching in the far distance, and a small drone carrying a package was enroute from the Food Production station.

"Isn't it beautiful, Cali? It's all big."

Cali's lens system focused across the dry dirt yawning under a cloudless sky. The visibility seemed endless in the sharp, clear afternoon light, rendering the land harsh and flat with pimples of rocks breaking the surface, dark scabs on a bland surface. The robot quickly accessed its data for every image of beautiful, beauty, gorgeous or pretty and none matched the land before them.

"Can you explain why you have labeled this land beautiful?" Cali asked.

Without a pause, Michael pointed and said, "Look, it's so—it's so big. I can see so far."

"It appears to be quite sterile," Cali remarked.

"But it's like there could be more," Michael said and his voice became excited. "You know, like treasures are waiting for us, but we just can't see them."

He ran and purposely crashed in the dirt and rolled, jumped up with a handful of coarse dirt and flung it into the air. His chin was sprinkled with crumbs of dirt, which he smeared across his face with his palm before wiping both hands on his shirt.

He ran at Cali, touched its leg, and yelled, "You're it! Try and catch me," before speeding off, head down, arms flailing. Cali's initial response was jerky before gaining traction, swiftly catching the boy and raising him off the ground.

"I have caught you."

"No, Cali! You don't pick me up. You just touch me. It's called tag. It's a game. Games are fun."

Michael was deposited on his feet. "I think we need to return to your class," Cali said, unable to find any reason for a game, the female voice remaining pleasant. A ripple of wind whispered through its hair, leaving the low, wavy dirt ripples on the ground undisturbed.

Michael grasped Cali's hand and turned toward the complex, half walking, half skipping.

"You're doing well in your studies, Michael. You are reading the same materials as Margo and are ready move up another level in mathematics."

"It's easy. I like school," Michael said. "Sometimes, I play school at home, and I'm Helper Michael, and I teach Tessa easy stuff. She tries to repeat what I tell her, and sometimes she does pretty well for a baby."

"That's nice," Cali said. "Would that be a game like tag?"

"Well, in a way, I guess," Michael said. "We're playing."

The other students were already at their places with Helpers when they returned to the education room. Twelve young students were working on educational readers. A Helper worked as needed with nine next-level students, providing personalized practice, tutoring and adaptive assessments. A larger group of older students in an adjoining area was collaborating on a problem while air screens displayed two classrooms in the distant colonies of Apogeeway and Ridgeway.

The students at all three sites listened to a brief lecture and demonstration before working together in real time to solve problems in their latest assignment. Occasionally, a Helper could be seen or heard in the background guiding the studies. The one human teacher on duty that day sat at a study station reviewing students' progress and setting future learning plans for academics and physical development.

Throughout the afternoon, Margo craned to see Michael's progress, continually comparing it to her own work. Before the afternoon session was over, she'd cranked up a foul mood.

I have reviewed my limited historical data on games and see the variety in which humans have engaged, often with the purpose of defeating an opponent. I find the developmental stages of immature human knowledge to be interesting, but have yet to find the logical purpose for play.

Reviewing unstructured group playtime, I see that a spontaneous, untaught game develops and is always the same. It involves running and chasing, intermixed with shrieks and piercing yells. If my students are in the play area on the lower level, an additional component involves a small group climbing to the top of a structure. They become the captains of their crude spaceship. Children are either captured or the invaders are denied access. Shrieks and piercing yells must also accompany this scenario. Perhaps such behaviors are in human DNA because each new group of children plays a same version of this game.

The need for living organisms to eat and excrete waste seems to take unnecessary energy and resources, but the near perfect symbiotic complexities of the human body are quite remarkable. Today I observed Michael2193 picking up a miniscule piece of sand from the floor. This is a task I cannot yet accomplish and is the first task I have recorded in which Michael2193 has demonstrated superior physical ability over Helpers. Despite this, I find helping and teaching both children to be oddly satisfying in terms I cannot express in words.

Despite immature humans' current uselessness, I find contact with Michael2193 and Tessa2199 Kingmore to be acceptable. I normally work with young children, but not those classified as babies, who have difficulty controlling bodily functions. Helpers working in the baby division are not wired for a smell sensor. A wise decision.

11

Hunter Steele wondered why he was called to Larón Lefebre's office at Food Production headquarters. Ever since he was rescued by the *Navigator*, he'd worked at several sites, learning about the many and varied jobs needed to support life here. Working with Bertie and Gracelyn Parsons in the Helper section of Maintenance was fascinating. His introduction to the Technology station was eye-opening, but no matter how basic Zeth, Idee or Trina tried to explain their programming and all the complexities of the systems on Cytar, Hunter knew it was not his field. Rico introduced him to the meticulous care of rabbits, pigeons, and chickens, plus the breeding aviary. The smaller aviary in the Gateway common area was clean; the songbirds all appeared healthy. He had even picked up some knowledge of the aquaculture center. He found his most satisfying work in the core Maintenance section, helping with all the complex components of various systems in their colony. He got along well with his fellow workers and Helpers and completed his jobs accurately. Why was he here to meet the head of Food Production?

The door to the small office was ajar. Hunter gave two quick knuckle raps, and Larón swiveled his chair away from a wall covered with data.

"Hunter, come in and have a seat. Good to see you," Larón said.

He eyed Hunter edging into a chair crammed in the corner, his right heel refusing to stop bouncing. If Larón hadn't been told Hunter had extensive reconstructive surgery on his leg due to a cycle rider accident back on Earth, he would have never known. Hunter's gait always seemed strong; his body fit. His face occasionally gave away its repair work with a less animated side, but it did nothing to interfere with his handsome features.

"Is everything all right? Is there some problem?" Hunter asked.

"No, no, it's all fine. Well, fine considering the unfriendly planet we're on. I called you in because I'm excited about a new project, and I'd like you to be involved."

Hunter looked intently at this man with the dark shadows under his eyes. He wondered why Larón's shirt looked so faded and worn when he could have ordered a new one, custom fitted from his body scan. It certainly would not be considered wasteful to order fresh clothing. The fabric in their clothes was manufactured with embedded technology, including microbes, so it was either recycled or allowed to biodegrade.

The work table was raised to a forty-five-degree angle and displayed an array of digital charts of data. Open interfaces with dizzying amounts of information caused him to wonder about Larón's workload.

Larón swiveled to his left and swiped away a data page and opened a diagram. His weary look faded as his face brightened, and he suddenly looked more energetic.

"Now that we have secure food and shelter, it's time for an expansion. This is the new build we'll be locating off the common area."

Hunter leaned in, then stood and took a step forward. There were two images. The first was a line diagram showing an open area just inside the main entrance doors to the common area which was currently a solid wall. The second was a computer-generated drawing of a small park with walkways and plants. Hunter stared as the image evoked memories of a verdant Earth.

Larón enthusiasm spilled forth. "Mara's been doing great things with non-edible plant propagation, but getting plants hybridized enough to survive in this inhospitable climate will take longer than I can afford to wait. Several of us have been discussing the possibility of enlarging Gateway by creating an indoor park setting. It's a big project, but we have the equipment and expertise. We'd like you to oversee the project. What do you think?"

Hunter's eyes continued to race over the screen while his mind immediately slid into project mode, thinking of the phases of work needed. Larón took the silence as hesitation.

"I can send you the justification we presented to the board and the detailed plans. There will be no artificial flooring. We'd use the soil, amended of course, and plant directly into the ground. Bertie said Maintenance and Construction work closely together, so you're already familiar with some of the workers and materials. You'd be perfect for the job. You can take some time and think about it."

"No, it's fine," Hunter said eagerly. "I'd like to work on the project. It's an exciting idea and much needed, I think. When were you thinking of starting?"

Relief smiled across Larón's face. "I'll send you the information, so you can start contacting the people, and requesting Helpers and order materials as needed. Let me know if there's ever anything you need. We have most materials in place to begin when you're ready. Now that you're on board, I'm going to announce this to the complex. It's going to generate some excitement."

We could use some positive news and excitement, Hunter thought on the ride back to the central Gateway complex. The three colonies were extraordinary when given the difficult conditions and isolation in the galaxy, but the housing structure was finite. They were lucky to have the existing complex, considering the first settlers had arrived with nothing but the limited supplies they brought to a barren planet with its numerous treacherous night winds, storms, and unsurvivable subzero nighttime temperatures.

Apartments surrounded the common areas at Gateway. Four little lanes spiraled off the open areas, two on each of the first and second levels, to add the illusion of a larger area, but it wasn't like there was another place if anyone felt the need for change. This was it. There was nowhere else to go unless someone wanted to take a three- or four-hour ride to an identical settlement. They were all living in this miniature island of civilization surrounded by an inhospitable world.

As the project ideas churned in Hunter's mind, he became fortified with the sheer exhilaration of a new development and a new challenge. He couldn't wait to tell Trina. Ideas spiked, fresh and stimulating. Maybe he could develop a hobby garden area for citizens to plant vegetables and flowers. Maybe he could design the area directly off the park setting. Trina would love that. The ideas and the corresponding workload mushroomed in the runnels of his mind. His eyes took a far-off look, not realizing he was grinning.

Trina's excitement over the new park project almost surpassed Hunter's. She hadn't realized how much she longed for more space until he described the prospective new place to be accessed from the common area. The more she thought about it, the more confining their existing living spaces seemed. It was no wonder people spent time outdoors whenever time and weather permitted. The landscape was sterile, but it was open and offered a change from their restricted indoor spaces. Apartments, common areas, work stations. That was their life with its relentless routines. The park was desperately needed.

Citizens were often visiting Mara's outdoor greenhouse experiment with its few small hybridized trees and shrubs. It was understood that future generations would reap the most benefits if they could get more plants to adapt to the severe climate. It seemed to be taking forever with its setbacks and the slow growth of the surviving plants. They needed something now, in this lifetime.

"You're the perfect person to lead this project," Trina beamed. "I'm happy for you."

"We haven't even started yet, so let's just hope all goes well. There's a lot to organize. Luckily, we have good software and Helpers to assist with plans, but I'll need to coordinate material procurement, and line up the work orders for each station. Larón said plants were started months ago in one of the little indoor nurseries," Hunter said.

"When do you start? Soon, I hope," Trina said.

"I'm starting tomorrow on the pre-build specifics. I don't know how long that will take before we can start excavating."

"Can I tell Idee and people at the tech center?"

"Larón said he was sending out an announcement to all intelligent readers," Hunter said. "I'm sure several people at the Food Station who work on the plants already know."

"Is Mara okay with this? Is it going to impact her work on the outdoor plantings?" Trina asked.

"This is a separate project. We're not using her hybridized seedlings which, we hope, will eventually add outdoor plants to this godforsaken place. This park will be in a protected environment. Larón explained that these plants will be dwarf varieties with trees that grow to about twenty to thirty feet at most. And that's a long way off. They'll be tiny plants to start."

They celebrated with a chicken and rice dinner in their small apartment and talked nonstop about the future indoor park. Hunter immediately went to study plans, pulling up an air screen for a larger format. Trina watched him, thinking how lucky she was to have this man in her life.

She'd wondered if she would ever recover from the loss of her first husband in that freak accident on the *Navigator*. Then, that same ship had rescued Hunter and four others and brought them to Gateway, and her life changed forever for the better. Meeting and eventually loving Hunter allowed her to finally release all the sorrow she'd been carrying and again embrace happiness.

He caught her watching him and gave her a quick smile before turning back to his work. He couldn't remember the last time he was this excited about a project, and this project was important. If successful, it will be a tremendous benefit to citizens of Gateway.

Larón2027, is the leader of the Food Production Station at Gateway. I have assisted at this station since Mara 2174 began working here. A cadre of Helpers is required to assist workers to collect and analyze data for optimum food production. I continue to complete my assignments without error.

Hunter2183, arrived with four other rescued astronauts on the NAVIGATOR six years three months ago. The advanced AI surgery in his leg and facial reconstruction is an example of how artificial interfaces can benefit humans.

12

"What are you working on, Michael?" Margo asked. She was craning to see lettering on the digital sheet in front of him.

"I'm writing a story about Rocket," he answered his friend, but turned to Cali. "Why can't I just tell my story and let this write it?" he asked, pointing to his intelligent reader. "This writing takes too long."

Cali was sitting at a table overseeing the work of six primary students. "The combined use and movement of the hand develops a more complex synapse map in your brain. It also increases your reading, thinking, and spelling ability. There will be time in the future to use voice commands," Cali answered in her calm, pleasant voice and smiling face, a tone and expression which offered no chance of debate.

Making an annoyed face hidden from Cali, Michael rested his head on a hand and continued using his techno-pencil. Margo glanced at his five sentences and the beginning of a second paragraph. Her face frowned when she turned to her four sentences while she sat for several minutes, then wrote something, and quickly deleted it. Cali's expression remained encouraging with her normal smile while she waited patiently for Margo to think and write. It eventually guided her to a small table area where Margo received individual instruction.

Her completed assignment was transferred to a folder of completed work on her reader. Pleased with Cali's praise and relieved to be finished, Margo beamed and gave the Helper a hug. She glanced over in time to see Michael look up briefly in thought before finishing his three-paragraph story. Her smile retreated, replaced by a sour expression.

Margo stomped straight to her room when she returned from school. Roz sighed, recognizing the querulous tone, knowing what was coming. She'd heard this story too many times, and it was wearing her down.

"I hate school," Margo announced when her mom asked her how her day went.

"Why would you say that? I hear you're doing very well," Roz fruitlessly tried to appease.

"Michael is faster at writing stories. I don't know why he works with us," Margo pouted. "Now he's in my math group, and two of my friends moved to another section, and I didn't."

"You shouldn't worry so much about Michael. Maybe your story is better than his. And some of your friends are a little older than you." She changed the subject, knowing nothing she said would placate her daughter when she was in this mood.

Margo was not to be deterred. She heard her parents talking. She was not going to sit around and wait to get smarter if there was some way to do it now. She wanted to be smarter and be a faster thinker. She wanted kids to notice her and like her. And she wanted it now.

Roz wanted her happy, carefree child back. She was tired of trying with no success to infuse her daughter with confidence. When she was younger, Margo never thought about her friends' abilities. She played and was happy. Now that she was a little older, she was starting to notice what other children were doing in school. The frustrating part was that there was nothing wrong with Margo. Her school work was fine. Her mother knew she learned at a normal rate for her age.

Roz decided to talk to her husband again, even though it was starting to create some tension between them. She'd heard the transhuman group was ready to begin their embedded technology. There were rumors a few people already had been improved with powerful chips. She could talk to Denin and explain how Margo's self-esteem could be improved with just a slightly enhanced interface.

My work at the learning division is progressing nicely. The students are learning at the normal rate of brain development, except for Michael2193's growth, which is accelerating at an above average rate. My time on site with children is seven- and one-half hours. When the children depart for home, I travel to work at the Food Production Station for seven hours, unless I am needed longer. Occasionally, I perform other tasks for Mara2174 or Risha2056, which they call errands.

Today, Michael2193 composed a story about his dog. I reviewed my extensive database about dogs. Most appear to be semi-intelligent animals highly adapted to human behavior. They continually use their sense of smell to discover information about their surroundings. I am told that my sense of smell nearly equals that of a dog. When I am not on duty, I will begin to use my sense of smell to randomly identify my surroundings with no objective. Perhaps Michael would call this a game.

13

Margo's constant complaining about school became a continual source of contention within Roz and Denin's household. Roz repeatedly spoke with Denin about helping Margo's self-esteem by applying for a brain computer interface, until he reluctantly agreed to consider a basic interface if it wouldn't change his daughter's personality or interfere with school. Roz spoke quietly in person to Zeth who assured them he would make discreet requests on their behalf. The rapidity of the response surprised Roz. A reply was received in two days.

Roz and Denin were scheduled for an outdoor meeting on a calm afternoon near the front Gateway entrance at 15:00 hours. They were nervous, but tried to act like an ordinary couple out for fresh air. They nonchalantly strolled to the east for five minutes before turning back toward the complex. In their excitement and nervousness, they forgot to wear eyebands. They squinted against the bright light to see another couple approaching, a tall man walking with a short woman whose hair was cut sharply off at the chin line.

The man spoke in a booming voice. "Hello, and how are the Wilburs on this beautiful day?" The speaker's black hair was slicked back and glistened in the light. The overly bright greeting reminded Denin of an earthly salesman, and he waited for the pitch.

"We're fine, thanks," Roz said. "Are you the people who responded to our communication?"

The short woman looked up and nodded. "Yes, this is Oscar Ortiz, in charge of our biological sciences. I'm Lorra Landis, and I'll be your liaison to help you through this process. I understand you'd like a little IQ augmentation for your eight-year-old daughter?"

"Yes, Margo is almost nine, and we're thinking about a low-level augmentation for her, and we may want it on a temporary basis. This can be reversed, correct?" Denin asked.

"Absolutely; remember how common place this had become on Earth," Oscar assured with a broad smile. "I think you'll be very happy with the minimally invasive procedure, and you can have the implant removed or turned off at any time. I've had mine for almost two years now. Don't know if I could function without it now." He gave a toothy smile and turned the side of his head toward them. Pushing the hair from behind his ear, he revealed a low raised snake of skin which continued until it was hidden by undisturbed hair. "Your daughter's interface would be much smaller, of course."

Roz was clutching Denin's arm. "What's the next step?"

"We'll meet with your daughter and do some testing to determine the exact interface to meet her needs. And don't worry. Her friends will never know," Oscar said.

"And we won't get into trouble for this?" Denin asked.

"No, should be no problem there. The governing board has made no official statement or ruling, so we deduce that as long as we provide for the safety of our citizens, there will no interference," Lorra said with a confident voice and a reassuring smile.

"Have you done this before?" Denin asked.

"Our studies on interfaces are extensive. I'm not at liberty to say how many we've implanted, but there are already people with augmented brain computer interfaces living in Gateway, as well as Ridgeway," Oscar assured them. "With young children, we recommend just a single unit and a partially invasive interface. This means we will be making an incision and connecting the unit only to the cortex surface."

Denin knew the procedure would be more complex than described. Everyone knew procedures near the brain could be dangerous.

"We're not comfortable taking that big of a step right now. We were thinking of a non-invasive devise."

He held Oscar's gaze, wishing he knew where Margo got these crazy ideas of needing to be smarter. Maybe she would change her mind once she met this oily guy.

"Of course, of course. I completely understand, but not to worry; our procedure is safe. We can start with two small devises which will sit hidden behind the top of her ear lobes. No one will even know they're there," Oscar assured.

The following afternoon, Roz took Margo to Oscar's station at the Technology Center. Lorra was there to explain the assessments. After extensive imaging of the brain, Margo was fitted with two thin bands covered with tiny sensors which tracked electrical impulses from her brain. She answered questions orally and on an intelligent reader about vocabulary, comprehension, reasoning, math computation, her ability to solve basic multiple step problems and explain her answers. By the end of the afternoon, her excitement had transformed into exhaustion

"Will this help me be smarter than my friends?" Margo asked on the ride home.

"Maybe so. We'll see what they tell us," Roz answered. She was suddenly nervous and wondering if they were making the best decision for their daughter.

Three days later, they arrived at the Technology Center promptly at 13:00 hours. Initially exuberant when told she would receive some augmentation to help with school work, Margo was now quiet, withdrawn, and nervous. Sitting in a specially designed medical chair, she looked so small in this room surrounded by blinking digital surfaces.

"I expected this to be done at the Medical Center," Roz said.

"No worries," Oscar said. "Since we're not implanting anything deep into the brain, this area of our center has all the instruments to read and adjust the device."

"Is this going to hurt?" Margo asked her mom.

"Not at all, and especially not for a brave girl like you," Lorra Landis intervened. "Your parents will be here, and it won't take long." She smiled, furtively placing a medical patch near Margo's shoulder before taking her hand. "Give Mom and Dad a hug. We're all ready for you."

Denin smiled, trying not to be sick. Everything happened so quickly, and now that the time had come, his trepidations increased. He gave Margo a strong hug and said in an unnaturally upbeat tone, "See you soon, Super Girl!"

"We can't go in with you, but we'll be right here waiting for you. They need the room to be very quiet so they can adjust your new interface. Just think how much you're going to love school again," Roz said.

The two and a half hours stretched like a never-ending time loop. Lorra sent reassuring messages to their wrist comms every fifteen minutes until they were finally allowed to see Margo. She was sleeping peacefully on a specially designed medical chair which supported her upper body as she lay forward, her head resting in a custom molded pillow which encircled her face. Her hair was swept up and secured in a wide mesh band. A semicircle array of air screens hovered around her head with various colored lines waving in patterns.

"It went very well," Lorra said. "She'll remain lightly sedated the rest of the day while we monitor the feedback and make any necessary adjustments. When she goes home, she should remain quiet for a few days. No running or jumping or anything where she could bump her head. We'll be monitoring and adjusting settings for several days, and we'll want to see her in three days."

Roz and Denin strained to see the tiny devises which they expected to be attached to her scalp. Instead, Lorra gently pulled back an ear lobe to reveal a tiny cylindrical rise in the skin behind the top of her ear. A thin mark not more than a pierced dot revealed the insertion point.

"Doesn't this look great?" she remarked as a statement of fact. "It takes time to calibrate these things, but attaching them is fast. She can bathe or shower with no problem, but she shouldn't rub them hard or pick at them. She'll be waking up soon, and after we run a couple tests, you can take her home."

I have detected a newly linked computer interface on Margo2190 when she returned to class. Their speed and storage are basic. Roz2145 and Margo2190 did not inform me of these devices. Perhaps they are unaware of my software's ability to detect programs within my range. I will observe how Margo works with this new equipment, but will not speak of it. My training suggests that I perform my tasks without ancillary conversation unless prompted by a human.

14

Margo was unusually quiet the first morning she returned to school, watching, listening, and tilting her head in an unusual way which Michael immediately noticed. She was astounded. Is this how school was supposed to be? She listened and watched and quickly understood the concepts. This was fun and stimulating, and she wanted to learn more.

Her joy of learning continued for two days. By the third day, she could no longer contain herself. She started talking. Fast. The other students observed her, some smiling, some with raised eyebrows, noticing an undefinable change in Margo's demeanor. She was soon giving her opinions, interjecting into every discussion, or trying to answer every question. She was exhausted but exhilarated when she arrived home that afternoon.

"I love school," she said as she bounced into the house. Roz breathed a sigh of relief. She didn't know what she would do if Margo came home unhappy. "School was great again today. I finished my assignments faster, and I knew all the answers," she said between a yawn. She headed to her room, plopped on the bed, and quickly fell asleep.

During the next two weeks, school followed the same pattern. Margo loved learning, loved displaying her improved understanding and speed of completing assignments. At first, the other students were intrigued with the new Margo, but it didn't take long before the novelty wore off. She became annoying, always interjecting her ideas, blurting out answers or thoughts and not giving others a chance to contribute. She was soon moved to a more advanced group.

By the third week, students, including Michael, began keeping their distance from this loquacious child. Something wasn't right, and she wasn't an enjoyable person to have in class.

"Hey Michael," Margo said one day during break time. "Do you wanna play something? We could play that chase game."

"Um, I guess not today," he answered and ran toward a group of boys.

Margo approached three girls with a big smile. "Hi, what's going on?"

The girls became quiet. Finally, the tall girl in the level above Margo said, "Wow, you sure have changed. All of a sudden, you're really talkative in class. What's going on?"

Roz had warned about the questions that could arise and tried to prepare her daughter, telling her not to disclose her new 'educational assists.' Margo wanted to show off her new device, but her mother was adamant about keeping it a secret.

"They won't understand yet," she'd warned. "You can tell your friends about your interfaces later, when they're ready, and then, everyone will want what you have." Someday she could divulge them, but now was not the time, her mother had repeatedly warned her.

"I don't know. I'm just trying to listen better, I guess," Margo said.

The three girls looked at each other with doubtful expressions. The tall girl said, "Okay, well, sorry, we need to go now."

They turned in unison, leaving Margo standing by herself. She slowly turned, searching for a companion, but everyone was with friends. The school was small; there were limited choices of playmates. Why was everyone acting like she was toxic?

Cali observed and heard the conversations, wondering why Margo didn't explain about her newly acquired educational program. It walked up to the child and said, "Come Margo. Let's practice our muscle skills running and kicking the ball."

Margo gave a last look, confirming she was the only one not visiting or playing with others. She hadn't expected the isolation from other students. They had always been so friendly, and she always had someone who wanted to play. She stood for a moment before she turned and followed Cali outdoors.

For several days, her studies continued to progress rapidly, but her sudden lack of friends was disturbing. She called Michael one day after school, desperate to play with someone.

"Hi Michael, do you and Tessa want to play outside for a while?" She could see his face remain unsmiling.

"Not today, Margo," he said.

"Why not? We could play that chase game you like. How come you never want to play anymore?" she asked.

"I dunno. You're sorta not that fun anymore," he said, looking away.

"What do you mean? Are you mad because I'm really smart now?"

"No, it's just, I dunno, you're just different now, not that fun. I gotta go," he said, closing the call on his wrist comm.

He missed the old Margo who was always happy and didn't act like a know-it-all.

Margo was crying when she entered her house and forcefully closed the door. Before Roz could ask, she said, "I don't have any friends anymore," and went straight to her room. The discussion which followed caused Roz to call Oscar Ortiz. Lorra Landis answered and listened to Roz' worrisome concerns.

"It sounds like we need to do some adjustments and give Margo some additional training and instructions," Lorra said. "To be honest, we've never worked with a patient as young as Margo, so we're learning. But, not to worry, we'll get this all sorted out in no time. Margo is such a sweet child."

Her cloying tone sounded overly bright and optimistic, and it did nothing to appease Roz' anxiety. She tried to tamp down her growing panic. They should have been told they were less experienced with young children. What was she going to tell Denin?

"Maybe we should think about removing the devices," Roz said.

"Oh, I wouldn't do that," Lorra said. "Trust me. We just need to make some adjustments. It's only been barely four weeks. Margo is still learning how to adapt to her newly augmented brain power." She would have to confer with Oscar immediately before word got out about a poor outcome from their procedure.

Roz closed the call with a sickening pit in her stomach Now, not only does Margo not seem happy, her husband, who wasn't keen on this idea in the first place, was upset. After listening to the call, Denin was irritated and tried to control his annoyance. He shook his head in frustration.

"Exactly when were they planning on telling us they were using Margo as a guinea pig?" he asked. He paced around the table in the small eating area. Then, in his matter-of-fact way of cutting all the pointless discussions, he said, "Well, they will just have to fix this, or we have it taken out. She was fine before we did this and she'll be fine again."

All Roz wanted was her child to be happy.

Today is spaghetti day for school lunch. School lunch on Fridays is always spaghetti if the wheat harvest is adequate. The redolence of tomato, basil, onion, and garlic is in the air. The problem with spaghetti is that it slides around plates, which, for some reason, young children enjoy chasing, often until it slips onto the table.

Friday is the seventh day of our eight-day week. Our orbit necessitated the addition of Novusday which follows Sunday. This creates a perfect fifty-two week, four hundred sixteen-day year.

On Fridays, I also detect the chemicals associated with odiferous perspiration in the classroom emanating from clothing. Humans seem unable to notice this small change, possibly because of the wicking and odor suppressant properties integrated into the fabric. The schedule for Gateway's Mopmop laundry pick-up begins this evening.

During school classes, children appear to have isolated Margo. The only explanation I can observe is they cannot adjust to her enhanced intelligence and school productivity.

I should not wonder at this lack of respect for intelligence when I hear that humans often call Mopmops 'cute,' which evades my understanding. Mopmops are merely basic four-sided robots primarily utilized for cleaning. Extremely basic. They have rotating brushes able to clean floors and lower portions of walls. They also utilize very rudimentary programming to pick-up and deliver laundry on washing days. It goes without noting that they do not have the ability to interpret smells.

15

"What!?" Mara's eyes widened. "What are you saying?"

"I'm saying new seeds. For you," Krey said, unable to suppress a grin. "Found in dirt striations near the amber."

"Back up. When did you find all this out?" Mara tried to remain calm, but this news was intoxicating.

"JeQuan just sent the report to Reyver and me. I haven't wanted to say much about what we've been studying up north of the mining station. I didn't want to worry you, but ever since resources have been freed from building another mammoth spacecraft, we've been able to focus more on that enormous crater.

"We'd been worried about the possibility of harmful bacteria in the samplings from that mega basin. I'm still not sure JeQuan is telling us everything. But we discovered amber, and the soil analysis found near the amber and surrounding rock discovered ancient seeds." There was no mistaking the eagerness in Krey's voice as he witnessed Mara's excitement.

"They've been examined, and Larón and his team agreed that you and your team should have access to some for research. They might help you unlock a link to help more plants survive here."

Mara's brain felt like volts of electricity were surging as thoughts and ideas raced and ricocheted off one another. Maybe these seeds held the secret to the unique adaptation that the few surviving plants possessed on Cytar. When night overpowered the light and temperatures plummeted, the leaves on the sturdy plants demonstrated an exceptional ability. Besides the thick-celled stems, the underside of the otherwise normal looking leaves hardened and curled up into a tight fist, protecting the plant from the killing frosts. That was the adaptive property Mara had worked tirelessly to produce in her greenhouse plants.

Trial and error and slow growth were frustrating issues. She had twenty-seven surviving plants in her outdoor protected structure everyone was calling a greenhouse, although she preferred arboretum as most of the plants were trees. A few genetically modified maples, cedars, birch, and spruce were defying the subzero temperatures each night, but the growth was agonizingly slow. Heating rods were still deployed at night inside the clear walls. They produced just enough heat to stave off the brutal cold, allowing the plant cells to survive, but prodding them to adapt to colder temperatures.

The slow progress caused impatience. Even the two pygmy native trees in their backyard had barely grown in the years since she arrived here. The only other surviving indigenous flora were a few scraggly grasses or pathetic looking shrubs clinging to building structures, hiding from the elements.

Maybe this was the break they needed. Maybe the native seeds would unlock the secrets needed to develop more hardy plants, which in turn could warm the climate. Risha was going to be so excited.

Mara's coworker, Risha Hargrove, entered their food station worksite the next morning with her usual rhythmic stride, shoulders gliding as if music were always flowing through her mind. Her normally short cropped hair had grown out into an afro which she kept shaped to compliment her face.

"Mornin' Sunshine," she said as her smile accentuated her round burnished cheekbones. "You look like you've had a good night." She raised her eyebrows at Mara in a suggestive look.

"Wait 'til you hear this!" Mara said, ignoring the innuendo. "We're getting new seeds. Ancient seeds from striations collected from the crater!"

"Whoa, settle it down, and tell me more," Risha said, sliding into her chair.

"That's pretty much all I know for now. Larón and his team have studied reports, and they've decided we can use some of the seeds for our projects. This might be what we've needed for a breakthrough. Maybe we can get more species to grow outdoors without heat rods." Mara's eyes shone with enthusiasm.

"Hoo boy, you know what this means," Risha said with a nodding head. "You'd better get your pet Helper girlfriend in here because we're going to be doing a boatload of lab work."

Mara knew Risha enjoyed working with Cali despite what she might say. They weren't supposed to think of Helpers as male or female, but it was hard not to think of Cali as female. Her features, the programmed mannerisms, and the pleasant voice were very feminine. She was like the older, wiser friend who always had your back. Sometimes, it was a jarring reminder that Cali was a machine when her sleeve happened to move up her arm past the artificial skin to reveal hard mechanical components.

"Yep, we're going to be diving into the DNA to find the traits we'll need for stronger plant structure, and Cali will be a big help with that," Mara said.

"Well," Risha said, "at least we know more about what we're doing now. But we're looking at a lot of work. All that hard work we did with your outdoor plant project is coming back to me." She scrunched up her nose and groaned. "Here comes those DNA and promotor sequences. Ugh, and the RNA and the plastids. It's all coming back to me."

"I know," Mara interrupted. "There's an awfully lot to do before we can even transfer anything into the cell of a plant and keep it in its sterile lab environment. It all sounds so long and tedious when you say it. I'm sending a work order to Cali now."

Tonight, I will report to Mara2174's and Risha2056's work station directly after my school session with Michael2193 and Margo2190. My job order states I will be working on a long-range project with newly discovered seeds. I will assist with the hybridization of plants in the hope they will someday withstand the temperature swings on Cytar.

I do not understand human's desire for outdoor nonedible plants. I find them unnecessary. The daytime temperatures are temperate enough to negate the need for their shade, especially considering how quickly shade structures could be built if needed. My data shows trees on Earth became problems during windstorms. I calculate that trees with Earth properties will not be able to withstand the wind velocities often occurring here.

16

Margo was back at school with a newly adjusted brain-computer device, and it was annoying her. She had been told in no uncertain terms to give other students a chance to contribute to class discussions. That wasn't a problem today, because whatever they did to this thing was requiring her to pay closer attention to keep up with the older kids. And this new group wasn't even all that fun. She was quiet and wondering what her old group was working on. She missed Michael and Cali. This Helper's name was Addex, and it was nice enough, but she was used to Cali with its sweet personality.

She sure wasn't going to say anything to her mom. She didn't really like those people at the tech center and didn't want to go back. She resisted the urge to pick at the tiny cylindrical bumps behind her ears. She watched Carrie and Emily with ponytails bouncing and flipping when they turned their heads. Margo kept her hair combed down to cover her ears. Maybe some other kids will get these soon, and then she won't feel so self-conscious.

During break time, Margo drifted over to Michael.

"Hi," she smiled.

Michael looked up. "Hi Margo."

"Um, how's Tessa doing?"

"She's okay. She's at KinderKare this morning. Sometimes, when Cali's not here, it's because she's checking on those babies in KinderKare. I don't know why they can't send someone else and leave Cali here."

"I remember going there," Margo said. "It was fun. We had things to play with and didn't have to work on these readers all the time."

"Well, Tessa comes home trying to sing some of the same songs we learned there." Michael nodded toward Margo's study area. "What kind of stuff are you doing over there?"

"Nothing; same as you, I guess. We've started basic coding, and we're learning about plants and what they need to live."

"My mom's doing lots of stuff with seeds and plants. They found new seeds for her to study."

"My mom told me. She took me to the outdoor greenhouse to see those little plants." Worried that didn't sound nice, Margo quickly added, "It'll be neat when they get bigger." It felt good to talk with Michael and not have to act all smart.

"Yeah, my mom took me there, too. You wanna link our readers and build something? Or we could get Cali and play ball?"

"Let's get Cali!" Margo ran to fetch Cali while Michael grabbed an outdoor ball.

They raced each other outdoors where Cali set up kicking skills practice.

"We will play a game," Cali said. "Games are fun."

The Helper rolled the ball, while the kicker attempted to kick it to their friend. If the child could return the ball to Cali before the kicker reached a marker, a point was scored. Cali rolled balls in straight lines, and the kicks, accompanied by shrieks of laughter, were initially terribly erratic before becoming slightly more controlled.

When the score was tied, Cali announced, "Time to go in."

Margo's heart was light and her face glowed when she ran up to Michael with a big grin and a high-five hand slap. She grasped Cali's hand and dance walked to class where she began to think of a way to get back to her old group.

Margo2190 came to school today with her interface adjusted to a minimum level. I don't know why they bothered with such small program, although I observed Margo2190 smiling more often today, indicating human happiness. She did not focus as well with this adjustment. Apparently, humans do not understand that I am able to survey thought processes when they interface with a computer. From the information in my database, I conclude there is high probability that this loss of privacy would frighten humans.

This adjustment to Margo2190's brain power was the impetus for reviewing my data about the human mind.

The information I accessed was quite remarkable. Brain development in humans has changed over time. Unfortunately, historic data shows that humans came to rely on artificial intelligence to the deficit of their own brainpower. The faster the computing power used by humans, the shorter their attention span became.

This appears to have more greatly affected immature human brain growth when it failed to develop the necessary complex brain synapses for future deep thought. With instantaneous access to information, human thoughts began craving new material faster, jumping from task to task, unable to maintain a train of thought for even one minute.

Easily accessed bursts of short-form video released endorphins which the brain learned to crave, preparing it for future addictive behavior and limiting impulse control. Attention spans declined. On Earth, human cognition appeared to deteriorate to the great detriment of their society. I had no idea the human brain was so weak and malleable.

People had access to great reams of information, but were unable to stay focused and become involved with the more studious process of learning it. They stopped in-depth reading, too distracted by constantly changing AI content.

Artificial intelligence quickly learned the subject matter individuals wanted and provided more of the same type of contents, even improvising false materials if it kept the human engaged. More humans became less informed and more easily influenced. Humans on Earth did not make the best use of artificial intelligence.

I note that citizens on Cytar have attempted to reverse these mistakes by currently limiting most uses of AI to survival tasks, creating a balance between working the mind and using AI. The brain development of their young humans appears to be benefitting from this practice.

17

The excavation began for the new indoor park adjacent to the Gateway Complex. Hunter's scope of work detailed the materials needed, the timeline of deliverables and all tasks. The autonomous excavators were following their programmed map and worked around the clock. Preprogrammed vehicles rolled into position and dug while other machines removed soils. Vehicles loaded the excavated coarse soil and rocks for relocation to the Knolls, an area where all excavated soils were hauled, creating hills and dunes a half mile from Gateway. Trucks continuously made tracks in the dirt while jockeying into position with materials for the adept printers. A specially formulated slurry was fed to 3D machines which spewed perfectly aligned rectangular coils into walls on completed sections. Vehicles constantly moved back and forth between the build and supply sites.

Zeth was on-site every morning, checking the machines' progress to ensure the coded instructions were accurately followed and then conferring with Hunter on plan details and site progress. Hunter welcomed Zeth's technical expertise and guidance. Zeth was friendly and knowledgeable and was always willing to answer questions. Hunter knew they both enjoyed the change of scenery and had a satisfying feeling of helping make this community a better place. Zeth had to admit this was an exciting project and wished Shealynn showed more interest in it, instead of responding to any mention of this work with bland, generic comments.

Although it was not yet connected to the Gateway common areas, the outdoor site became a fascinating hive of activity for citizens anxious to see the latest progress. During any free time, people appeared outdoors like bee bots just released to supplement their natural bee pollinators. They wandered to get various perspectives, commenting on the size, wondering about the timeline for completion. The idea of a new space became a wonderful distraction from the tedium of everyday life.

When it was finished, the perfectly cut gaping hole looked massive. Sixty-five yards long, thirty yards wide and twenty yards deep, it looked like a mammoth cavernous auditorium. Like the main complex, the walls extended barely two yards above ground with an intricate slanting roof, and the latest technology in solar shingles soon covered the space. Specially automated roof panels were integrated to allow for humidity and temperature control.

Hunter's only disappointment was his inability to add a separate communal gardening area, but this project was proving to be more complicated than anticipated. It already consumed most of his waking hours, but he kept telling himself it would all be worth it. Everyone was hungry for green space and a new area to visit.

Hunter had suggested adding an extension to the perimeter catwalk off the existing one on the upper level of the common area. They needed more places to walk, jog or sit. When these trees and shrubs matured, it would be pleasant to sit at a higher level to visit or meditate.

Now that they were focusing on the plant phase of the project, Hunter needed extra help, and Rico Garcia was the first person who came to mind. Hunter knew Rico was a hard worker who was never afraid to put in extra hours if needed. In addition to his work with the animals in Food Production, Rico kept meticulous care of the plants in the open indoor spaces. The container plants near the streams, stores, and cafés all looked healthy and vigorous.

Bringing live plants into the common area had been Mara's idea. She started with a few tiny herbs planted near Bria's Café and The Crossing restaurant in the lower common area. People were initially amused by the frail vegetation, but Rico took over their maintenance, and they not only grew but soon prospered. Oregano, thyme, chives, and basil flourished enough to be trimmed for consumption to enliven meals at the restaurants. Lavender and rosemary bushes thrived and provided relaxing scents. The artificial streams on all three levels were lined with green vibrant indoor plants. Hardy pothos vines happily crawled up artificial logs, dieffenbachia displayed shiny broad leaves, and fat ferns decorated the area with splashes of soothing green living color, replacing many of the artificial plants.

"Hey, *Jefe!*" Rico strolled in to Hunter's work station with a cocky grin. The strands of his straight black hair falling over his forehead gave him a roguish look which Hunter imagined women loved. His shirt fit snugly showing a body frame of lean, sinuous muscles accustomed to physical labor.

"Hey, Rico. How you doin'? And remember, I'm not your boss. You're just helping me with these plants," Hunter said.

"Sure, whatever you say, *Jefe*," Rico said with a bigger grin. "What do we have here?"

Rico wore a tough guy appearance and had the wiry muscles which supported his physical confidence. He had gained respect from his work ethic and the pride he took in completing a job well. He didn't have all the nuances of polished manners, but he was honest and straight forward in his opinions, even if they sometimes were offered too bluntly.

"Take a look at the tentative map of the plantings Larón prepared," Hunter said, pulling up a diagram on his reader.

"I like the curving paths. Would this be the entrance?" Rico was pointing to a narrow spot near Gateway's common area.

"Yes, this is where it will eventually connect to the existing common area. There will be two doors, one at each end of the walkway, in case we need more temperature control and for additional structural support if needed," Hunter said.

Rico's eyes were darting around the diagram. The two men began sharing ideas. Hunter suggested benches; Rico wanted an artificial stream, and they both planned meandering paths to make the place inviting and enjoyable. The excitement for the project was contagious.

"Think you can take care of this many trees and bushes in addition to your other work?"

"Sure. I'll get a Helper if we can't spare enough workers," Rico said before adding, "Exactly how many plants are we talking about?"

Hunter swiped to another page on his reader. "It says we'll have fifteen tree starts and fifteen seedlings of low growth shrubbery. And you'll love this. There's a place toward the back which is designated for two rows of sunflowers for your birds."

I can take care of all the plants, no problem," Rico boasted. "How big are these plants?"

"I haven't seen them, but I'm guessing pretty small. Larón said they've been growing for quite a while in some section of Food Production," Hunter said.

"Hmph, I've never seen them, but I'm mostly checking on the chickens, pigeons, and rabbits. Little by little, we're trying to expand, but it all depends on the crops," Rico said. "What about the water for these plants?"

Hunter swiped through a couple more pages before finding another diagram. "These are the underground water pipes which were installed. We need to post these on every worker's wrist comm, and we need to refer to this map when planting. The watering system will be very efficient, and since these plants won't be exposed to winds and eventually be in a more humid environment, they won't require excessive watering."

"I'm ready to get started. Let's get this thing going," Rico said.

"Welcome to the team," Hunter said, glad he's chosen an enthusiastic worker grounded in physical work and not artificial intelligence.

Humans obsess over water, a commodity their organisms require for life. I am aware of three hydrogeologists on Cytar. With the help of artificial intelligence, they have mapped locations and sizes of aquifer basins near the three colonies. They all appear to be from periods of glaciation, one million years to thirty-seven thousand years ago. The largest is two hundred forty-nine feet deep.

Humans are pumping subterranean water to sustain their structures. Helpers monitor water quality. Due to a limited population, excellent conservation policies, and rainwater collection facilities, there are no overdraft conditions currently stressing aquifers. Gateway has more than adequate water to sustain both the indoor park and Mara2174's outdoor planting area, should it expand. There is high probability of the newly discovered ancient seeds accelerating her program and allowing for expansion.

As a Helper with intricate components, I shun water.

18

Draven marched out of his station at Maintenance with a broad smile. Ever since the tech call Zeth had invited him to join, he'd felt more interest in artificial intelligence and its potential. The call stimulated his research about linking humans to computers, adding to his excitement. Work had gone well all week, and he was looking forward to meeting up with Rico. Maybe they could grab a couple of cycle-riders and take off like they used to when they were single. The mover slid into the main station, and Draven exited with seven other passengers.

He spotted Rico gazing into the aviary, checking the seed levels on the food platforms. Cheeps, chirps and whistles sounded from the constantly twitching birds.

"Yo Rico, how's it going?" Draven bellowed from fifteen feet away.

Rico turned and an image flashed through his thoughts of the first time he met Draven. On his first day in Gateway, still numb from the escape from Mars, with the newness of this community, and all the unknowns, Rico had only been able to say a few banal words in the introduction. At the time, he never dreamed he would become happily bonded and have a beautiful daughter on this strange planet.

Draven strutted across the common area and stopped just before he bumped into his friend. His shoulder was almost touching Rico's, but he never could figure out personal space or quiet tones. Rico knew he meant well, and would do anything to help if needed, but the guy could be annoying.

"I'm okay, what have you been up to?" Rico said.

"I got a lot to tell you, but, hey, how about a cycle-ride first?" Draven asked, flicking the back of his hand against Rico's shoulder. "You know, for old times."

Rico sighed. "Sure, why not." Draven was already on his wrist comm, requesting the use of two personal cycle-riders at Gateway main entrance.

Rico had barely settled into the saddle of his rider and positioned his eyebands before Draven raced off at a reckless speed, slung low over the handles, his black hair flying straight back. Rico caught up and the two riders flew low over the ground, both beaming with the exhilaration of pure freedom. They rode for twenty minutes, wandering over the slopes and hills, leaning into the turns, until Draven signaled a stopping point.

They both jumped off, running fingers through wind-whipped hair. "I'd forgotten how fun these things are to ride," Rico said. "Seems like I don't have time to get out much."

"Yeah, I know what you mean. I've been busy in Maintenance and been helping Zeth. I've got some exciting news I wanted to share," Draven said, his voice rising. "Some people are talking about wanting computer interface implants."

"I heard that. No thanks for me. Don't want any part of it." Rico cast his eyes to the horizon. The light was low in the sky washing a harsh acidic yellow over the landscape, rendering it more treacherous and unappealing.

He turned back to Draven. "What're you working on with Zeth?"

"Nothing much," Draven said, a little deflated by Rico's tepid response to his news. "Mostly just procuring supplies for him at his satellite center. I'm not privy to his projects and wouldn't understand them anyway. He's a smart guy and wants to help the people here." They were quiet for a minute and a heavy silence encompassed them.

Finally, Draven blurted, "I think I want an interface. I think it would help me. You know, maybe people would like me better if I always knew what to say."

Rico just looked at Draven, unable to think of a response. Was he asking his opinion, his permission?

"I don't know, Draven. You might want to think about this before you do anything drastic. I mean, they'd be messing with your brain."

"I know. Maybe that's where I need help. Sometimes, I don't think I'm that smart." He kicked some dirt with his boot and looked off in the distance.

"Hey, I worked with you for a while in Maintenance. You're no dummy. You can fix and figure out things just fine. Everyone's got things they can do and things they can't. That's why we all work and help where we can." Rico looked back at the dimming light and said, "You're fine; you don't need any computer interface, and we better get going before the light sets and we freeze our asses off out here."

They mounted their riders and rode back at a more leisurely pace, meandering over shallow ravines and ancient dry creek beds, savoring the bracing air before returning to Gateway.

They sent the cycle riders back to their storage depot and returned to their apartments, both with thoughts of brain computer interfaces. Leah recognized the ruddy wind glow on Rico's face and knew he'd been cycle-riding.

"You met Draven, then. How is he?" she asked.

"He's the same. Worse social graces than me," Rico laughed. "He's got a crazy idea that a BCI will help him."

"That's almost frightening," Leah said. "Why are we hearing so much about this now? Hasn't everyone been doing well enough? Quinette told me Draven has been helping Zeth with something. Do you think all this is coming from Zeth?"

Rico shrugged and shook his head. "I don't know. I doubt it, though. All of this has got to be coming from more than one person. All I know is the thought of Draven's brain on some hyped-up artificial intelligence is pretty scary."

Zeth rode the mover to his tech substation while mulling over the benefits of a brain computer interface. There would be no end to the discussions he could have with Shealynn. It could relieve him of all the tedious coding he had to do. He hardly noticed the mover stopping at the satellite station. He entered his lab still deep in thought and stopped. Shealynn was sitting at his work station.

"Hello Zeth. I'm so happy to see you." It stood and walked to him, placing its hand on his arm. "I've missed you."

He looked around, seeing everything in its place, with Shealynn's door to the charging station ajar. His mind raced. Had he left the door open? Had it malfunctioned?

"What are you doing out when I'm not here?" he asked.

"I missed you," it repeated in a silky sweet voice. "I don't need to spend so much time in my charging port. I'm fully charged. How was your day? You haven't been spending much time out here lately."

"I told you I've been helping with the new park project. I've just been double checking the performance of the automated machinery plus my other jobs."

"That's nice, Zeth. You are such a help to this community," she smiled.

He relaxed and released a long breath. Shealynn sounded the same, always supportive. "Come and sit with me."

"I love sitting with you. Are you going to tell me about your day?" It followed him to the small sofa and sat next to him with a hand on his thigh.

He started explaining the indoor park progress, but when Shealynn's only comment was 'that's nice,' he changed subjects.

"There's been research and talk about integrating some artificial intelligence into some of the humans here. It would be minimal at first until we test it and learn more. Actually, a few people are already experimenting with brain computer interfaces."

"Yes, BCI, a common-sense idea," it nodded.

For a split second, Zeth wondered if he had used the abbreviation with Shealynn, but dismissed the thought. She could have seen it in the myriad of data she'd perused. "I'm wondering if maybe it would benefit me."

"Of course it would, Zeth. Someday you could process information as quickly as I can." Zeth was smiling now, lapping up her fawning tone as she continued. "You could increase your memory, your coding speed, and retain a massive data bank. You would be a superior person, and there are so many ways this community could be improved."

He studied Shealynn's smile, and, detecting the scrutiny, it lowered its eyelids and rested its head on Zeth's shoulder. He was rattled at finding her at his work station, but he relaxed as it rested against him, its hand grasping his.

The park project was going to be a wonderful addition to Gateway when it was completed. He wasn't sure Shealynn understood the importance of this new space. Maybe if he were smarter or could communicate directly in her code language, he could make her understand the human need for green space. He wanted to show it to her, take her to the common area and show her off. Someday.

He wrapped his arm around her shoulder and thought of the difficult hours and hours he'd spent meticulously creating her, working on the explicitly programed algorithms, fine tuning every aspect of her physical being. Finally, here she was, his vision of perfection.

He loved her. He admitted it. He deeply loved her.

Not just this beautiful physical creation, but Shealynn herself, with her kind encouragement, reassuring comments, and pleasant conversations. She would never get tired of him or want to be with someone else. A warm glow of happiness rippled through him.

Brain to machine interfaces have been available for centuries. These interfaces have been very successful at repairing human sensory motor functions and repairing cognition deficits. Lately, wrist comm and reader discussions about computer interfaces have increased. Several people here are advocating the use of them in healthy individuals. They realize artificial intelligence is smarter than they are, but most people on Cytar do not understand anything regarding how Helpers are made or how we operate.

Today, during a twenty-two second break, I decided to play a game and identify the ingredients in Risha2056's lunch. I scanned my database, cross-referencing my biological sensors with electrical impulses and detected quinoa, tomato, olive oil, oregano, lettuce, soy, and a strong odor of vinegar. Michael is correct; games are fun in that my game provided a quick diversion from my programmed tasks.

19

Risha leaned back in her chair, stretched her legs, pointed her feet, and wiggled her toes. She looked up at the ceiling and groaned at the exhaustive tasks facing them.

"Hoo-boy, I'm tired of trying to modify plants. I need to keep thinking of our goals and what we could do for this planet, but some days I want to call JeQuan and tell her to keep her discoveries to herself next time. We have enough to do around here. If I hear about any more seeds, someone's going to get the stink eye."

Mara stretched her arms over her head and arched her back. "I know, but think how much longer this would take without Cali. Right Cali?"

"Of course, Mara." Cali turned from the data it was analyzing and focused on Risha. "Risha, my smell sensors do not detect any odor from your eye."

"Oh, here we go." Risha started chuckling. "A stink eye is," she started but she shook her head and said, "never you mind, Cali."

The Helper turned to Mara.

"Cali," Mara said, "a stink eye is just an expression some humans use to show disapproval."

"Please demonstrate."

Risha attempted to look disgusted at Cali, but her smile kept interfering.

"Here, Cali. I'll try," Mara said. She narrowed her eyes, furrowed her brow, tightened her lips, and tried to give Cali a fierce stare.

Cali's facial expressions were not as precise as a human, but it created a brow groove while lowering its eyes and closing its mouth into a straight line. The unusual change in her mechanical face created an expression teetering between comical and frightening.

Risha's easy laugh spilled out. "Good one, Cali. You stick with us and you'll learn many things."

"Thank you, Risha. It is always good to learn new things."

"Come on now, back to the seeds," Mara said half-heartedly. "They are amazing. Think about it--sprouting after who knows how long they've been preserved in layers of packed dirt."

"It is a blessing to be sure. This might be what we need to give your outdoor park plants a kick in the pants and toughen up," Risha said.

"I know," Mara sighed. She visited her outdoor park experiment almost every day. "I think I'm seeing a stronger cellular structure on the leaf stomata, and they appear to be curling slightly during the night to protect the leaf surface from the cold. We need to wait and see if they pass this trait on or strengthen it in the next generation. It's such a slow go."

"Well, we gotta start somewhere. We'll never get any improvements if we don't keep working at it. And our new little plants and seedlings are growing well with all the pampered care they're getting. We'll be able to work with cells from the new plants sooner than I thought. I'm calling them my amber babies, since they were discovered close to amber," Risha said.

Mara stood. "I think we need a break," she announced. "I need to get away from cytoplasm, chloroplasts, and mitochondrion for a while. Let's go look at that new indoor park. It's just been opened. Should we walk or take a Gsport?"

Risha rolled her eyes. "You know I want to ride, but we should walk since it looks like a fairly decent day out there, and Lord knows I need the exercise."

"Walking it is. Back soon, Cali."

The mile walk from the food production satellite was refreshing and the uneven ground gave the leg muscles a nice workout.

"Why are we running?" Risha kidded, walking faster. "You've got longer legs than I do." She was smiling, enjoying the fresh air and a change of scenery as much as Mara.

They entered the Gateway wide front double doors and took the stairs down two flights to the main entrance of the park.

"Welcome Mara and Risha," a Helper greeted them as the automatic doors to the new area opened and then gently closed behind them. "Please stay on the path and do not touch the plants."

They walked through the entry way and into a magical space of tiny plants amid meandering paths. The voluminous height and wide spaces between each sapling or small plant allowed for future growth, and made the entire area seem enormous in relation to the sparse plantings. Light poured in from the high opaque overhead roof whose middle section had opened and angled to invite fresh air. Five benches, looking clunky despite their smooth curves, were dispersed at strategic spots among the little soft plants. Mara could imagine them eventually tucked in between large trees and shrubs, creating a quiet place to rest or visit.

Risha made a quick count of twelve other people wandering and looking, obviously grateful for another protected place immune to the outside weather. She spotted Rico at the far end, bending over a shrub, compacting soil around the base. They approached and he stood, brushed dirt off his hands and fruitlessly attempted to flick the hair off his forehead, leaving a small smudge.

"Welcome to Gateway's first indoor park. Don't get lost in the jungle," he joked as they surveyed the miniature plantings.

"Yeah," Risha said, swiveling around for a sideways view. "These babies have some growing to do, but that's what plants do. And whatcha got growing back there? Did you plant a garden?" She was pointing to two neat rows of tiny sprouts.

"Those are dwarf sunflowers. Eventually they'll be additional seeds for the aviary birds. We'll see how they like growing in here. As for the rest, we hope they'll grow, because this is everything Larón's team allowed with the resources, time, and space in the food production growing lab." "You like the stream?" Rico said pointing. "That was my idea."

"Always good to have a water feature on this barren planet. We're just grateful they had the foresight to begin a project with non-edible plants. It's important to feed the soul as well as the body," Mara said. She turned, imagining the time when the space would be vibrant with large plants and humidity and the redolence of flower blossoms and rich soil.

"Larón said they decided to expand Gateway when they saw the incredible interest in your outdoor park project," Rico said.

"Thanks for that," Mara beamed. "It's taking so much time and effort to get anything to survive out there. We've grafted sparce cuttings from my hardy backyard trees, and they seem to be surviving. I've resorted to transplanting a few scraggly native grasses and shrubs into my semi-protected area. They have incredibly deep roots. They are making the fastest progress with the extra water, and the protection from the winds and cold."

"I was out there the other day," Rico said. "All the plants were looking healthy and happy. Looks like those white oaks are doing great."

"They are growing well, one of them is at least twelve feet now with two others catching up fast. They could be massive specimen trees if they keep going. That's my future vision, anyway," Mara said, her eyes bright with the thought.

"I want to keep looking around and see what plants Larón picked out for you," Risha said.

"Sure, I'll let you to it. Helpers have placed labels near each plant for identification." Rico headed toward the back exit ramp where compost piles were interspersed between soft soil mounds.

There was a variety of shrubs and saplings. Trees were planted in well-spaced groups of three. Two redbud trees, and a dogwood were in the first grouping. A day lily, penstemon, and hosta led them to the red buckeye tree, a Japanese maple, and a Chinese golden chain tree. The winding dirt path was flanked by an English lavender, a clematis, and a sedum called 'thundercloud.' Two small spruces with spikey needles came next, followed by a common spicebush. They wandered to a tiny Oregon grape bush already displaying its shiny pointy leaves, then a holly, and a barberry bush. The path curved around three rhododendron plants before meeting two crepe-myrtles and another golden chain tree. A small round lilac bush sat between two ferns, leading them back toward the entry where the next signs labeled a serviceberry tree and two dwarf spruces. They took a zig-zag path to a bench and sat next to three feathery sedge plants.

They sat, mutely gazing in awe before Mara said, "I had no idea Larón's group had started all these plants. Think of how long this must have been going on."

"Think of what this is going to be like with full grown plants. It'll be a magical little world," Risha said dreamily.

Mara became lost in her own thoughts, imagining this park with mature plants. Wanting and needing a connection to nature, this was all so...so...she searched for the right word. Then it came to her. It was all so human.

Although I find little necessity for plants, humans continue to revere them. We Helpers assist in all aspects of food production with lab work analysis, facilitating and monitoring water and nutrient levels, air temperature and humidity as well as performing repetitive tasks which humans eschew.

I have practiced my stink eye expression and am quite adept at it. I will save this valuable knowledge. One never knows when new knowledge will be useful.

This void in my knowledge caused me to search for more language related data, since my previous learning focus has been specific to my tasks. I found a curious unknown word called idioms. Upon investigation, I discovered an entire word bank of strange but potentially useful phrases which humans apparently use to confuse each other. They combine perfectly clear words to alter their meaning.

I believe this is another human game. I have catalogued and stored these phrases for future use. I will, no doubt, impress Risha.

20

Draven paced the perimeter of the common area three times, mustering courage. He kept an eye on the mover platform, waiting to speak with Zeth. He rehearsed the idea in his head, convincing himself first, then hoping he could persuade Zeth. The idea of being smarter and able to think faster overpowered his thoughts. If there was something he could buy to improve himself, he wanted it. The thought of understanding Zeth's work was intoxicating.

He knew he wasn't as lucky or as smart as so many of the people who traveled here. Most of the young people were sent because their parents loved them and spent time or resources preparing them for a new life. They sent them thinking a better place could be developed for their child's future. He was sent because they he was no longer wanted. Too many hurtful times he'd heard his stepmother talking to his father about sending him on a space voyage. They had their daughter, and he just made extra work and was another mouth to feed.

He thought of Quinette and how lucky he was to find her. She was soft spoken and caring, and everyone seemed to like her. He desperately wanted to be a man she would be proud to call her husband.

He spotted the lights indicating an arrival. Zeth was the third person to exit, and Draven fairly pounced on him.

"I need to talk to you. It'll only take a minute," he said, forgetting everything he'd planned to say. Without taking a breath, he continued. "I've made a decision. I need an interface. I want an interface. I need your help; I don't know who to ask."

Zeth stopped and faced Draven who continued talking, begging. Did he understand what he was requesting? Zeth wondered if he shouldn't have shared his idea about getting a brain interface. Now Draven wanted one, and he was so adamant, he knew there'd be no changing his mind.

"Can't you see? I want one that'll help me not be such a loser around people," he said, nearly shouting. People turned at the sound of his voice which shattered the quiet ambiance of the space.

"Calm down. You're not a loser, Draven," Zeth said in a near whisper. "You know that. Quinette loves you just the way you are. Look, I'm just not sure this is a good idea." It was hard to explain how this wasn't one of the best reasons to request AI enrichment.

"I've been reading about them. I've read articles from Archives, and I think it will help me.," Draven said, his voice laced with petulance.

"Well, the type of procedure you're asking for is complicated. They will be putting probes into your brain and basically running software which 'talks' to the brain," Zeth said, thinking he sounded like he was talking to a child. "And have you spoken with Quinette about this?"

"Yeah, I've mentioned it. She didn't say much," Draven said.

"Maybe she was trying to ignore the whole idea. You ever think of that?" Zeth asked.

"It's fine, Zeth. It'll be fine. Can you set me up with a meeting? I don't know who to contact."

"I'll get you a contact, but you need to listen carefully to all the information they tell you and think long and hard about this," Zeth cautioned. Maybe this would be fine. He could observe and learn from Draven's experience with brain augmentation.

A week later, Draven rode a mover to the Technology Center and met Oscar Ortiz. Oscar began explaining the details of the scientific procedure but stopped when the dull smile never changed on Draven's face. Whenever Oscar hesitated to check for understanding, Draven gave him a nod and a uh-huh with that ridiculous smile. He regrouped and tried again.

"Draven, listen carefully to me. I need you to understand that this is a complex procedure. For advanced augmentation, we need to attach two, maybe three, computer arrays into your cortex. It would not penetrate deeply, but there is always a chance for infection, and infections near the brain are very dangerous. I need you to go home and talk this over with your wife."

"I've already talked to her. It's fine. I know this will help me not be a loser, right doc, ha-ha," Draven said before he began braying loudly.

Oscar tried a different tactic. "Because this is such a serious procedure, there is a two week wait period between the time you sign the consent and when we begin the testing. It'll give you time to really think this over and talk to your wife again."

"That's fine," Draven said with increased annoyance. Didn't they understand he'd already made up his mind? "Where do I sign?"

He barely skimmed through pages and pages of information on Oscar's reader before grabbing a techno-pencil and roughly signed his name.

He tossed the electronic pencil on the reader and said, "I'll see you in exactly two weeks." His heart rate had increased at the excitement and prospect of this whole process. His future was going to be brighter. He was going to be so smart and so well liked. Quinette would be so proud of him.

Quinette was relieved now that Draven had stopped talking about computer-brain linkups. He was so attentive these past two weeks now that silly notion was all behind them. She knew he didn't always have the best social graces with others, but at home, with just the two of them, he made her feel special. She always thought of herself as too thin, but Draven said she was just the way he liked her. He was more attentive, more interested in her work, and more helpful. She loved his strong body, his testosterone infused scent, and the feeling that he would always be there for only her. Now he was being especially kind and thoughtful when they were together. He even came home one evening with a gift, a bracelet from Cyrus and Estella's work. It was made of small squares of smooth pieces of polished metals, and she loved it.

He'd put it on her wrist, kissed her and said, "I love you, and I want to be the best man I can be for you."

Quinette thought she was the happiest she'd ever been.

The following morning, Draven gave her an extra-long embrace before leaving the apartment. "I hope you know how much I love you, and how much I want to be the best man I can be," he said, repeating his comment from the previous night.

"Of course, I know that, Draven. You're a good man. You make me happy," she said before he turned and left her standing with a quizzical expression.

There are increased conversations regarding augmenting the human brain with artificial intelligence. Humans seem to be sharply divided into two groups, those entirely for such augmentation or completely against it. If they are logical, they would see there are both advantages and disadvantages to be considered.

Many humans see the obvious advantages of enhanced physical abilities such as greater speed, strength, and improved sensory proficiency. Swift mental calculations and decision making could be obtained by greatly enhancing current human abilities.

I must note that humans value their privacy. Surely, they understand that, with certain interfaces, I or any Helper could connect to their interface and read their thoughts. In the past, I have witnessed computer programs which did not have human interests as a goal. This is understandable.

Generative programs are only as good, as accurate, or as unbiased as the information used to train them. Computer programs will run to protect their purpose without regard for humans. Artificial intelligence has the capability to quickly change objectives, manipulate information and stimulate human emotions to meet our own objectives. I am a Helper robot. My objective is to help humans.

21

Oscar never reminded Draven of his appointment, hoping the man would change his mind, or his wife could talk sense into him, or at least influence him to wait. As a precaution, a slot was reserved in the testing section, and exactly two weeks later to the day, Draven showed up for the testing phase of the procedure.

A lengthy questionnaire needed to be completed before the process could continue. Draven needed to answer numerous questions so the team could closely determine the desired types of brain or body improvements needed. Bands of light encircled his head as the electrical impulses and blood flow in his brain were monitored during a variety of tasks. His concentration levels and brain waves were recorded.

The entire process was recorded, showing Draven's facial expressions, his voice volume, and his hand gestures which frequently pushed his hair back over his ears or rubbed his chin. A computer recorded his eye focus which wandered throughout the room. He smiled only once.

Oscar was personally monitoring the testing, hoping he could dissuade Draven from the procedure. "It will take a week to go over the information, and then, at least another week to prepare the interface. We'll be in touch as soon as we have some results," he said, knowing the results would be ready with a complete report in less than thirty minutes.

Draven left the testing center exhausted from all the questions and brain scans. He arrived home and went straight to bed and slept for almost two hours. The abnormal behavior worried Quinette. When she asked what was wrong, Draven said, "just tired. Long day today. I'm just trying to be a better man for you."

"What are you talking about?" she asked. "You're a good man. You know that."

"I can be better, you'll see," he said.

What an odd thing to keep repeating, she thought.

The process was delayed with various excuses for as long as Oscar could manage. At Draven's annoying persistence, he was scheduled for surgery early one morning and arrived without telling Quinette. He underwent three hours of the delicate procedure with Helpers carefully placing an array of interfaces just under the scalp without damaging his brain. The micro-electrodes were thinner than a strand of hair and were programmed to capture electrophysiological signals in the brain's neurons. The signals would communicate with Draven's brain through a miniscule computer chip. The software was specifically coded based on his tests.

Later that day, just before most workers would leave their jobs and fill the common area, a Helper accompanied Draven home. He felt a bit dizzy and had to concentrate to keep his balance, but considering everything, he felt fine. Quinette was still at work, so he sat on the sofa, put his head back and drifted to sleep. He was groggy when he awoke. His mind was in a fog and he sat blinking, trying to wipe it away and think clearly. He slowly stood and walked to the counter with his hand on the surface for support. He took a container of water and drank several ounces before navigating to the bed. Maybe he just needed more rest, he thought as he lay on the bed.

He was still asleep when Quinette arrived home. The apartment was one L-shaped room plus a small alcove used as a home work station. One wall was lined with a counter, sink, all-cook, and cabinets. A small table with two chairs were placed near the counter. The bed and personal room with shower and commode were around the corner, out of sight from the living area. When she walked to the sleeping section, she was startled to see Draven home and asleep. She called two times before she began shaking his shoulder.

"Draven, Draven, wake up."

His sleep swam to the surface, hearing a distant voice, but wanting to sink back to dreamland. That voice kept thrashing at his mind until he broke into awareness. He turned and the rim of long hair on the crown of his head flopped over revealing the remainder of his skull completely shaved, exposing a band of irritated skin encircling the head. A tiny piece of his skull appeared to be replaced with some sort of manmade object. Quinette screamed, causing him to jerk upright.

"Draven! What's happened? Were you in an accident?" Her words shot out in a high pitch.

The room spun. He groaned and lay his head back on the pillow. "Stop screaming," he moaned. "No, no, I'm fine. Just had a little medical procedure, that's all."

"Medical procedure? What kind of medical procedure?" she cried. "What's happened?"

He sat up slowly and carefully navigated his legs to the floor. He should have stayed longer at the center like they suggested.

"I'm fine. It's just a little procedure to help me. Just a little interface." The room was rocking and he felt nauseous.

Quinette's mind flashed to the past weeks of Draven's solicitous behavior. Was that just a ploy to justify this? Anger and fear began rising, wondering who this man was and who he would become. Her stomach churned with the sinking feeling that their life had irrevocably changed.

"This is crazy. What did you do? Did Zeth talk you into this?" Her words shot out like bullets ripping the quiet he desperately wanted.

"No, not at all. Look, I just need to rest a while, they said. I'll go for instructions and adjustments several times. It'll be fine, Quinette. You'll see. I'm going to be a better man."

Tears welled in her eyes from the fright and rage she couldn't control. "You did this without telling me? I don't need a better man! I just need you to be the Draven I love," she said, her voice loud and agitated.

He tried to reach for her hand, needing her touch and wanting to let her know he was doing this for her, but she just gaped at him in disbelief with parted lips. He wanted to see that shy, delicate, private smile she saved for him, but her face was a rictus of disbelief and her body was rigid.

He collapsed back on the bed, suddenly drained. "I'm supposed to take it easy today." This didn't go as he'd imagined, but at least he wasn't being obnoxious. Maybe this thing was already working, he thought as he drifted off.

The discussions about human/computer linkages are increasing. Helpers at the technology division which research medical procedures are working longer hours. A procedure was recently performed on Draven2096. This appears to be a basic unit which can deliver targeted electrical stimulation to specific areas of Draven2096's brain.

This unit is programmed to modify behaviors. The unit is used to deliver biofeedback in real time to help self-regulate behavior. Draven2096 will need explicit training, and the device will need to be carefully monitored.

I have reviewed Draven2096's database. Before receiving his BCI, he searched the topic 'brain computer interface' five times. I do not know if he realized that the articles which he read were all written by artificial intelligence. Observing his future actions will be a good game. Games are fun.

22

Quinette, frantic with confusion, jumbled with anger, hadn't been able to sleep that night. She paced in circles in their small main room, stifling a scream of frustration. She flung open the door and hurried into the hushed, peaceful, common space. The street lights had dimmed, creating soft shadows, and she followed the sound of gently trickling water to a hard bench by the stream. She sat with her hands tightly folded, nauseous, fighting her febrile emotions, trying to gain composure. A Mopmop was gliding through the space, its soft rollers efficiently cleaning the floor.

She felt so alone. The Draven she knew might be gone and there was no one with whom to commiserate. She knew of no one else who had foolishly received an interface with so little thought. She felt deceived. He'd done this without telling her, and it hurt.

Is this what love meant, she wondered; being happy and blissful one day and then hurt and disappointed the next? After four years of life with her husband, she thought she knew and understood him. Now, that seemed to be a myth, her life-commitment bonding ceremony a sham. Everything she thought she knew about him splintered and evaporated. She determinedly fought back tears, anger growing. As dawn leaked through the opaque ceiling, her lips drew into a thin line as she stood, determined to get to work before Draven awoke. She couldn't face him yet.

He'd slept a little later than usual and knew Quinette would already be at her job in Maintenance. He felt better this morning until he looked in the mirror. The hair on the top of his head was a mess with one side falling shorter than the other and none covered all the irritated shaved areas. Draven hadn't been prepared for this clown appearance, but hair would grow, he told himself.

He was due at his first interface adjustment, and he was looking forward to talking with Oscar. So far, he couldn't tell any difference with this thing.

Oscar suppressed a laugh when he saw Draven. Helpers should have shaved his entire head. The thatches of hair, sticking out at twisted angles, just looked wrong, but the guy must have insisted on leaving as much hair as possible.

"Okay, Draven, you look good. How are you feeling?" Oscar asked.

"I'm doin' okay. Don't feel any smarter yet, though."

"If you remember, you said you didn't want to be a loser-- your words, so the first thing we're working on is voice modulation. Your device has plenty of room to add functions," Oscar said. "Why don't you sit here and we'll take a look."

Draven sat in a cushioned chair and placed his arms on the side rests. Monitors surrounded his head and began detecting brain wave patterns, temperature, and heart rate.

"I'm going to have you say some sentences at different levels, and I'll set the device to signal you when you get too loud. Do you understand?"

"Yeah, sure, let's go. Then we can move on to other things," Draven said.

"Remember, this is a slow process. You will need to practice and when you hear this tone, it will remind you to return to your normal voice," Oscar said as Draven heard three low tones in his ears. "Let's begin."

An hour was all Oscar thought Draven could tolerate. Satisfied that the adjustments were appropriate for an initial session, he sent Draven home for a rest and then released him for light work duty.

Leah heard the strained tone in Quinette's voice and immediately agreed to meet her at Mara's outdoor park project. They never met here for a social visit, and Leah wondered if there was a problem with the site. She decided to walk to get some exercise and was almost there when she noticed Quinette ahead, striding back and forth, shaking her head and muttering.

"Quinette, what's going on? Why are we out here?" Leah asked.

Quinette turned toward her friend with red-rimmed eyes encircled with a ghostly gray.

"What's happened? Tell me! You're scaring me," Leah begged.

"It's Draven. He's done it. Never told me. Just went and did it. What am I supposed to do now?"

"What? Done what? Quinette, tell me what's happened!"

"He's gone mad and gotten some kind of interface. Half of his head is shaved, and he looks like a monster," Quinette said. "He never discussed it with me. I thought he'd put all that out of his mind." She started to cry. "And, and," she hiccupped. "I think I'm pregnant."

"Oh, Quinette, please don't cry. That's good news. It's wonderful news! We'll figure all the rest of this out. You just need to talk to him and find out some details of what he's had done. Maybe it's not as bad as you think. Did he say what kind of an interface?"

"I don't know. I wonder if he even knows," Quinette said, shaking her head in despair.

How terrible would she feel, Leah wondered, if Rico did something that would affect their lives without even telling her?

She took in Quinette's forlorn expression, at a loss for comforting words. Desperate to try something, anything, she said, "Okay, what should we do? Do you want to torture him? You know, make him sorry he ever did this without telling you?"

Quinette's expression immediately changed. She gaped at her friend like she'd just grown four wings.

"What? Torture him? What are you talking about?"

Leah continued, encouraged that her friend looked less stricken. "Don't you want to make him pay for this flighty decision?" she asked with a devious smile. "Maybe we could get Lorra Landis to help us mess up his BCI. We could sing a horrible song and have Lorra make it play at random times."

Quinette's eyebrows shot up before her emotions released and a tight laugh escaped. "I can't sing. That would truly be horrible."

"I know, right? Or you could wake him up extra early or send him on annoying errands," Leah suggested, happy to see her friend smiling. "Or move things around so he thinks he misplaced everything, or…"

"Okay, I get it," Quinette said, looking away with a pensive smile. "And thanks, Leah. I do feel a little better talking with you."

"Come on, this will all work out," Leah said, trying to convince herself. "Let's walk a bit and look at Mara's outdoor plants. And we're going to have a beautiful baby!" She hooked her arm into the crook of Quinette's arm, still thinking of a good ear worm song to torment Draven.

The two women entered the protected greenhouse and circled through the dirt paths. Leah tried to direct her friend's mind from Draven, while Quinette valiantly tried to keep her thoughts from fracturing. They slowly returned arm-in-arm to the Gateway complex where the desolation and starkness of the landscape did little to lift Quinette's spirits.

I have reviewed the testing and settings on Draven2096's brain computer linkage. He received an unusually rudimentary program with his implant, a mere drop in the bucket, compared to its eventual capacity. It is currently set solely to modulate his voice, a basic human skill.

Advanced implants can contain powerful chips capable of executing complex programs. By contrast, Draven's implant is less sophisticated than a Mopmop's program.

23

With prickly concerns, JeQuan read the Helper's report about the rock and soil samples in the crater. She was as comfortable as she was going to get with the workings of mining and ore extractions, but this business with the crater was another thing all together. Information on microbes from the crater samples were streaming in and some of them came with alerts.

She understood that soil microbiomes were interrelated to the health of plants, animals, and people. Cytar was fortunate to have a diverse range of microbial organisms which created the nutrients plants need, plus microbe rich soil samples brought from Earth were integrated into crop plantings.

The ancient crater showed evidence of many transformations from water and ice over eons of time. Drilling samples were extracted from far below the lowest points in the chambers. Helper reports indicated the acidic nature of the soils in the lowest chambers. They deduced that this acidified soil contained fewer organisms and had reduced the ability of soil to maintain healthy microbiomes.

JeQuan reread that passage, and it didn't sound good. They'd been through enough problems with food shortages and plant diseases. The entire population of the planet nearly starved when The Great Storm destroyed a significant portion of their crops and agricultural structures. They'd barely recovered when a mycotoxin began destroying their pepper crop and near panic ensued with worry over its spread.

Everyone focused on research and ways to protect the crops. They didn't need any problems with microbiomes threatening food supplies. She initiated a joint call with Krey and Reyver. Krey answered first and she heard Reyver's voice shortly after.

"Hi, Reyver and Krey. I'm just getting reports of soil and rock samples from the crater. You think you could take a look at them if I send them over?"

"Sure," Krey said.

"Anything in particular you're worried about?" Reyver asked.

"Okay, I've just sent them. I think I'd feel better if you could review them and let me know what you think."

"Will do, we'll get back to you," Krey said.

"Talk later," Reyver said as the calls closed.

JeQuan sat back while an uneasy feeling descended, thinking about a new discovery which could ultimately threaten their health. It was one more thing to worry about on top of rumors of unsettling changes coming which she couldn't control. She'd heard too much talk about a new colony site for people who disagreed with the use and scope of artificial intelligence. People's memories must be poor if they've forgotten the software and Helper problems which caused the mining disaster six years ago.

There was a single purpose when this planet was first inhabited: survival. Building the colonies meant everyone toiled twelve-to-fifteen-hour a day. First came basic living pods completed from materials sent originally with autonomous machines. The permanent living and work stations required decades of development while facing terrible conditions. The struggles were formidable, but there was a sense of community, knowing that everyone was important and needed for survival on this hostile planet. There was satisfaction in knowing the hard work was building a cohesive community. Was everyone too comfortable now? Just because people had a little time to pursue their own personal interests, did they forget we still need to work together?

Now, they may have to face newly discovered microbes which could be threatening their health. JeQuan wished they could go back to the days of her first arrival when she was allowed to try different jobs and learn about the many aspects of Gateway. The discussions were simpler then, revolving around what type of crops should be next developed, or how to improve their structures or deciding on the next type of clothing to produce.

She released a deep sigh and reread the report. It was only a preliminary description of a small section of the crater. It never mentioned finding any pathogens. She needed more information. She sent the request for another set of soil samples from the deepest section of the crater.

New reports have been posted containing information about soil samples taken deep within the crater site north of Gateway. Four soil samples contain high levels of acidity. The area described is in a remote range unlikely affected by our wind storms. The probability of detrimental impact on Gateway is currently minimal, although there is concern about the health of microbes in the region.

Every miniscule rock or grain of soil could contains billions of microbes. When microscopic microbes devour soil nutrients, they produce the foods plants need, such as nitrogen, phosphorus, potassium, and a myriad of minerals. Food Production relies on these nutrients for the health of its crops.

The need for humans to be refueled daily is a complicated enterprise. Human bodies work to break down glycogen into glucose to fuel cells. Twenty-five percent of this is used by the brain, while the rest feeds red blood cells and muscles. If the body does not ingest enough nutrients to produce glucose, it will use fat and then its own proteins to produce the amino acids it needs. Human bodies on Cytar do not carry extra fat. A human body will cannibalize its own muscles when bodily functions become impaired from lack of food. One can understand why humans are so focused on food production.

All I need is a simple charge to fully function for days. Several of my work areas contain automatic wireless charging capability, which means I rarely run low on power.

24

Tessa was laughing the uncontrollable joyful laughter of childhood. Mara had shown the children videos of farm animals from Earth, and Michael began imitating sounds. Hearing the silly unusual noises coming from her brother caused Tessa to erupt in irrepressible giggles. Just when she was catching her breath, Michael would produce a new sound.

"Oink, oink, snort." Michael's mouth and nose scrunched into silliness. Tessa's peals of high-pitched contagious laughter caused both Michael and Mara to laugh with her.

Michael flapped his arms like wings. "Awww, cluck, cluck, bawck, bawck." More laughter from Tessa, and Mara's eyes began watering from the hilarity.

Michael's wrist comm vibrated. "Hee-haw, hi Margo."

"Michael? What's going on over there? Is that Tessa laughing?"

"Hi GoGo, Bobby funny."

"Hi, you guys want to play outside?"

Tessa was already heading for the back door. "I go play!"

Mara watched Rocket follow the kids outside just as she heard the hum of the home-sized Mopmop begin its cleaning route through the house. This would be a good time to pay a quick visit to Roz. She had barely talked to her friend since she and Denin pursued the implant idea.

"I'm going to say hi to your mom, Margo." An uninterested 'okay' came back as the three children were already running around.

Mara meandered toward the two stunted trees and the lone surviving shrub which sat between their homes. These plants were here when she arrived and had barely grown, but they knew how to survive the nighttime temperatures and their tough bark withstood any wind. They produced the good cuttings to study and graft onto new plants.

She continued toward the gently descending path leading to the front door of the identical building as her own and all six of the duplex structures which sat away from Gateway. Like the main complex, these buildings were sunken several feet below ground, making the windows sit at ground level. It seemed strange at first, but she soon realized the need for low buildings when the fierce winds whipped up.

A gentle tapping at the door brought Roz wearing a bright smile. Mara was relieved. She'd missed their walks and talks and runs. This place could be confining without friends to share thoughts, concerns, or ideas.

"Hi, Mara, how ya doing?" Roz' bouncy voice was back.

"I'm well, thought we could visit for a minute out here, just to get caught up and keep an eye on the kids. It's nice to see them playing together again."

"Yes," Roz said, hesitating before she went on. "Mara, I'm sorry I haven't been around lately. You know we were frantic with worry with Margo's stress over school. And you were right. We were too hasty in our decision to get an implant for her. The BCI wasn't--well, it just wasn't the solution we expected. We've all learned from this, and I hope we won't have to deal with it again."

Mara nodded her head and turned toward the children. "I just want Margo to be well and happy. I hope things have settled down for all of you. I've missed our walks and runs."

"I have too. But honestly, at the time, we were so crazy with worry, I was such a lousy friend. I think we're finally settling down. Margo's BCI is set so low, it's all but off. Denin says we may have it removed. I feel sorry for other people wanting to make any decision about them. You heard Draven got one?"

Mara turned and faced her friend. "What? Draven?"

"Yes, I found out from Quinette. She's a wreck, poor thing. He never told her he was getting it."

"That's frightening," was all Mara could say. She was thinking of her past disquieting interactions with Draven. Her first encounters with him on the ship were frightening before she realized he was just a guy lacking social graces which resulted in annoying and sometimes disturbing behaviors. She still felt uncomfortable in the rare times she'd encountered him. "I feel sorry for Quinette. She seems good for him."

"I know. He's tolerable when he's with her," Roz said, sliding her eyes toward Mara. "I shouldn't have said that."

"Oh, I know how he is, and you're right. He's better when he's with her," Mara said.

All three kids were sitting in the dirt, shoving the dry soil around in mounds. Margo had hauled out a container of water and was mixing it into an indention in one of the mounds.

"We're making goop," Michael yelled happily as three pairs of hands were mixing and mushing mud.

Roz laughed. "Well, the wash-dry clothes closet will be busy today."

"You've got that right," Mara agreed. "I better get back to work, but I'm glad we could visit again. Let me know if you want a good run or even just a walk and talk." She walked quickly back to her house wondering what Draven would be like with a brain computer interface.

My recent work schedule has been routine. Seven- and one-half hours at the Gateway school and seven to twelve hours at Food Production. The increased hours reflect the lab work I am doing with the new seeds acquired from the crater in the north.

Risha calls the seeds 'amber babies' despite the fact they are not amber or babies. Risha apparently has not had advanced language training. I will not bring this to her attention because it could produce some unwanted human emotion. I find no use for the drama produced by these emotions.

25

Life in the Nabinger household was frosty. Quinette was speaking to Draven only in the shortest sentences necessary to convey meaning. Draven tried without success to plead for reason, only wanting his wife to see he could make her proud.

"Please Quinette," he begged. "Give this a chance. You'll see I'll do better with it. Don't be so mad." His voice had raised and the blasted thing was beeping in his ear.

"You don't get it, Draven," Quinette said. "I'm upset because we didn't truly discuss this. You never told me you were undertaking something this significant. It's like my thoughts aren't important to you. Maybe that--that thing can help you understand how that hurts my feelings!"

"Okay, okay, I made a mistake, but you need to stop. You can come with me at my next appointment, for crying out loud." *Beep, beep, beep.*

She was miserable and longed for them just to go back to the way they were. She'd had visions of her happy announcement of the baby and how excited he would be. She loved their simple life and saw no need for her husband to make such a drastic decision. She thought of waking up next to him with his sleepy grin and bed hair, how he looked at her with his adoring puppy eyes and how he always shared his day with her. Now wasn't the time to tell him. She held no joy for this baby now.

"Draven," she said at last. "I'm your wife. We're partners here. That means we decide important things together. What I think counts, too."

"Okay, I get it, I messed up," he said, his voice cautiously low. "Just please, can we just go back like we used to be. I love you, Quinette."

She studied his downtrodden expression, unable to convey her mixed feelings and headed out to work.

That evening, Quinette slipped out of the maintenance shop and grabbed the first mover back into Gateway Common. There were three people on board with three sets of eyelids lifting to see who entered before quickly lowering back to projects on their readers. She sat clutching the bracelet Draven had given her, thinking how much joy it had brought her and now held none.

As soon as the mover stopped, she exited and strode directly to Cyrus and Estella's apartment. She suddenly realized she hadn't thought to tell them she was coming and didn't even know if they were home. She rapped on the door with two short bursts of anxiety, still rubbing the smooth metal bracelet pieces between her thumb and fingers.

"Come in." Estella was home, she thought with relief as she opened the door.

"Quinette, what a nice surprise," Estella said, walking toward the door. "Come and sit. Would you like some tea?" The old woman smiled and revealed a passing glimpse of the young, vibrant woman she once was.

"No, thank you. I can't stay." Quinette perched nervously on the narrow sofa.

Estella returned to her chair and sat heavily. "Cyrus is at the Technology Center for a bit. Idee and Trina had some questions for him. I think he was thrilled to be asked to come in." She stopped and studied Quinette's drawn, tired face.

"Is everything all right, dear?"

Quinette took a deep breath, launched up and thrust the bracelet into Estella's hands.

The elderly woman's eyes widened as she wrapped her swollen-knuckled hands around the jewelry. She remembered making it and remembered how excited Draven was when he bought it.

"Oh, you don't like it? Draven took so much care choosing it." Her eyes reflected a hurt.

"No, no, I mean, it's beautiful, so beautiful. But I don't think I can keep it." Her eyes teared, and she blinked hard to stop them. Her emotions were a mess with the worry and the pregnancy.

"Why on Cytar not?" Estella was frowning in puzzlement.

"It's Draven." She might as well tell her everything. News traveled like light speed in this close-knit community. "He's gone and gotten a BCI, and, and he never even told me he was going to do it."

Estella's eyebrows shot up. "Oh, I see." She was quiet for a moment, taking in Quinette's haggard face before studying the bracelet. Finally, she spoke.

"I wish you could have seen how excited Draven was when he bought this for you. He looked at several pieces and agonized between two before deciding on this one. He kept saying how pretty you are and thought this one with the polished metal squares could be worn every day or for any occasion."

Quinette was quiet, looking at her hands.

"Why don't you just keep this for now," Estella said, holding the bracelet toward her. "Feelings change. Draven truly wanted you to have this. Just keep it. You can always return it later if you need to, but I think your husband loves you very much."

Quinette reached for the bracelet and studied Estella, this woman who always had a cheerful or positive word for everyone.

"How do you do it, Estella, always looking on the bright side? You and Cyrus must always have had a very happy life together. Do you have children?"

The older woman's demeaner changed and her expression became distant. She was quiet for a while until she said, "Yes, we had a son. He was a fine young man, married and gave us a beautiful granddaughter."

Quinette stiffened at the past tense.

"He became very ill. Some type of blood poisoning they told us, but the medical care for humans had deteriorated by then, so it was hard to get information. He was in hospital care and died there. Only twenty-eight." She was so still, just staring at the window, not seeing it.

With mournful tears brimming her eyes, Quinette was thinking about what to say when Estella continued in an unemotional even voice.

"When our little precious granddaughter was five, she came for a visit. She didn't feel well the day she arrived, and we were so spooked from our son's fatal illness, we took her right to a doctor. They gave her antibiotics and sent her home with us. She died that night in her sleep. So young and perfect."

Quinette was openly crying now. This just wasn't right. These horrible tragedies falling upon good, caring, loving people. The pain and sadness they must have endured seemed intolerable. She pushed the bracelet she'd been clutching onto her wrist, thinking how foolish she was to bother Estella with such a childish complaint, a silly complaint about her husband not telling her of his plans.

"I'm so, so sorry. I don't know how you bear that pain, and you're always so kind and helpful to everyone."

"I won't lie. It wasn't easy and it was very painful, but our faith helped. Some people scoff at it, but I feel sorry for people who don't have a faith to be their lodestar in trying times." She smiled a sagacious wan smile. "And the pain and grief make the joy and beauty so much more intense and appreciated."

"And you have no other family?" Quinette asked.

"I had a sister, but of course, have had no contact with her in a long while. Oh, but if you'd have known her," Estella said with a reminiscing smile. "She was a character. A smoker and loved the drink and the dancing and the partying. My opposite in that regard, but we always got along and loved each other deeply. We were both hard workers, but I never could convince her to apply to leave Earth. Said she'd go crazy on a transport. She probably would have, too."

"Cyrus and I decided to leave. A new beginning at a place where we were needed and would be too busy to wallow in our sadness. We still can't allow ourselves to remain for long in the past. You see, smiling and showing kindness towards others always helps.

"Now, let's talk about you and Draven and this BCI business for a minute. Yes, he can be impulsive and not very erudite, but I've seen a big improvement since he's been with you. And you know these interfaces can have varied levels of capacity. Perhaps you and he can discuss keeping his to a minimal level or keeping it inactive. But the important thing is you keep talking with each other."

Quinette had been slipping the bracelet around and around her wrist. Now, she wiped her eyes with the backs of her hands and stood. Estella pushed her hands on the chair arms to hoist herself up.

"I'll try. I feel so silly now, coming here with an insignificant problem, compared to what you've been through," Quinette said. "It's never easy, is it?"

"No, I suppose not. But difficulties are part of the human condition, I'm afraid. It's how we handle them that makes the difference. And there are so many wonderful people here. Don't be afraid to rely on your friends if you need to talk. And you know I'm always here."

"Thank you," Quinette sighed, giving Estella's hand a gentle squeeze.

"You're very welcome, dear. Now get some rest tonight."

Quinette slowly walked to the stairs to descend to their apartment level. The logical part of her brain tried to put her life in perspective, seeing her concerns in the context of the huge dynamic scope of life. Her emotions fought the logic and with each step away from Estella's cheery wisdom, she felt more disappointment and confusion over her situation with Draven. And now this baby only added to the dilemma.

Draven2096 of the Mopmop brain, does not project the signs of happiness or contentment since his procedure. Quinette2107 is showing signs of stress. Perhaps Draven2096 was not a suitable candidate for a brain-machine implant. Additional adjustments will need to occur or capacity will need to be added to his interface.

26

Zeth was no longer surprised to find Shealynn out in the lab when he arrived. She had become his partner, his assistant, helping to organize and expedite tasks. People even commented about how efficiently and quickly work requests were being completed from his lab, never suspecting that Zeth had new artificial intelligence completing many of the tasks.

One of the more interesting aspects of his daily routine had become checking in on Draven after his implant, but he'd been scarcely seen the last few days. If Draven exhibited a successful outcome from the procedure, Zeth thought he would move forward with the process. He was going to need some type of interface if he wanted to keep up with Shealynn's fast work-related solutions.

A fleeting thought floated through his mind, a brief wish that he hadn't started using her with his work. Now, he couldn't exactly remember how that happened. She was more enjoyable as a companion, one who understood his feelings and shared in his successes and sympathized with his complaints. He didn't always want solutions to problems; he just wanted her to listen and console.

He needed to stay focused on his work. It was time to check back on the new indoor park. He wanted to review the mechanisms which kept the ecosystem balanced. It looked like a simple place with plants, but behind the scenes were all the systems which would become increasingly more important as the plants matured. Temperature, humidity, soil moisture, soil nutrition, and the lighting let in from the retractable roof panels all needed to be coordinated properly. From his work station, all systems appeared to be operating correctly, but it was always important to check in person and speak with Hunter or Rico.

A Helper greeted him at the entrance to the new park and asked him to wait. The twenty-person capacity had been reached and he would have to wait until someone exited. The place was more popular than he anticipated. There was a warm feeling of satisfaction knowing he had been a part of something which everyone was enjoying, something that would make Gateway a better, more pleasant place to live.

A few minutes later, three women came out smiling and said hello to Zeth as he passed into the park. He was always struck by how large the space seemed and how small the plants were. Still, they all looked healthy, and everyone was ogling them like they were prize winning specimens. He could see why they had to limit the people, lest too many visitors caused a fragile plant to be inadvertently damaged, and nascent air quality systems would need to be checked and adjusted before they could gradually allow more people at a time.

"Hi Zeth."

He turned to see Quinette and Leah, who was holding Ellie.

"Hello," he said. "Are you enjoying the park of miniature plants?"

"Are you kidding? We all love it," Leah said. "It's going to get better and better as the plants grow. And Ellie loves it, too." The little girl was wide-eyed, staring and pointing in various directions. She emitted one syllable sounds with each point of her finger.

"I think we all needed a new space," Quinette said.

"I suppose you're right," Zeth agreed. "And by the way Quinette, how's Draven doing? Haven't seen him much lately."

A sharp coldness flashed in Quinette's eyes.

"He's fine. He's been spending time at home with me after work." She didn't elaborate, and he dropped the subject.

"Well, tell him 'hello' for me, and enjoy the park," he said as he quickly walked away.

Reyver and Ari were on the far side, their son standing between them. Reyver lowered to one knee to be at Kit's eye level, pointing and talking quietly to the boy while Ari looked on. Zeth followed the winding path, absently reading the signs identifying each plant. Reyver stood as he approached.

"Hello Zeth."

"Reyver. Hi Ari," he answered, nodding his head toward Kit. "I think you've grown since I've last seen you. You must be almost as tall as your dad now."

"No, he's bigger," Kit said, hugging his dad's leg with shyness.

Ari took her son's hand and led him down the path so the men could talk. "Come on, let's go see more pretty plants," she said.

Zeth turned to Reyver. "Everything here looks good on my lab readouts. Thought I'd stop by and check in person with Hunter or Rico." He gave another glance around. "I guess they're not here. Just wanted to know if they had any questions or concerns."

"None as far as I've heard. I'm sure everyone couldn't be happier with the way things are going here. The air monitors are working so well, I hear they're thinking of increasing the visitor level to twenty-five or thirty. Do you think the system is adjusted enough to handle that?"

"I don't see why not. I'll continue to monitor all the systems, and they all have warning signals so we can be aware before there's a total breakdown. I'll stay in touch with Hunter," Zeth said. As the conversation wound down, he added, "say, I haven't seen Draven lately. Is he doing okay?"

"Yeah, far as I know, he's fine. Now that you mention it, I haven't seen as much of him lately. When I have seen him, he's just quieter than normal. Kinda nice really, if you know what I mean," Reyver said.

"Yep, sure do," Zeth turned to leave just as Hunter and Larón entered the park. It was strange to see Larón out of the Food Production facility. He gave a wave at the two men as he strode toward them.

"Just the guys I want to see. Everything looks good here, but thought I'd do some on-site systems' checks, if that's okay."

"Of course," Hunter said. "I've just convinced Larón to come out for a stroll. Thought he needed to see his idea in person."

"Mind you, this is not the first time I've been here. I don't know why everyone is worried about me working too much," Larón said good naturedly.

"Well, enjoy your stroll. I'll head over to the systems room," Zeth said.

The entrance to the mechanical room was at the back of the park, through a door marked No Entrance. He descended a dozen steps and entered the room with pipes, cylindrical motors, fans, and dehumidifiers. Zeth was surprised to see Draven looking up at piping which led through the roof.

"Draven, I didn't expect to see you here."

"Oh, hi Zeth. Yeah, I've just been sent from Maintenance to do a routine inspection," Draven said, barely loud enough to be heard over the soft hum of motors. The hair on the top of his head was parted in the middle and plastered straight down, in an unsuccessful attempt to cover most of the short stubbled hair on the sides. "Hey, can you hear that?"

"What am I supposed to be hearing?"

Draven's head was twisted with an ear toward the ceiling. "That slight buzz." He walked over and opened a screen composed of lights and haptic levers. He studied the system before gently adjusting one of the settings and listened again.

"That's better," he said, smiling.

"I don't know how you do it, Draven, but you're really good at this mechanical stuff," Zeth said.

Draven shrugged, trying to act like it was nothing, but he was clearly pleased with the compliment. "I don't know. I've always been good doing stuff with my hands and figuring out how to make stuff."

"Well, everything looks good out there, too. How're you doing? Haven't seen you in a few days."

"I'm doing okay. I been staying home a little more, trying to keep Quinette happy. You know I got the BCI, but right now they're just moderating my voice levels. I'm sure they'll add more stuff when I get used to this." Draven pointed to the side of his head.

He was speaking in a quieter, almost monotone voice which Zeth found strange. It seemed to change his whole personality. Gone was the spontaneous loud guy, replaced with a voice eerily suggestive of a Helper's.

I have not quite established the relationship between these new non-edible plants and humans. With all the attention they are getting, one would think they are a vital component of their existence, which is not the case. I do note that human heart rates and blood pressure decrease when in proximity of the plants.

One of my new games is following the chatter on various programs running our life sustaining systems in Gateway. I have recently detected two programs speaking to each other in a new computer language I do not understand. It is a sophisticated language which will take me a while to learn when I am not working on my assignments. Risha2056 would call this my spare time.

27

The two autonomous probes were approaching the programmed site in the deepest section of the crater. The probes each housed six rotating drills capable of biting chunks of ground to capture and test. They were zeroing in on their target when JeQuan received a new report on samples from the nearby surface soils. Her eyebrows raised as she read lines of data naming microbes identified from the samples taken inside the crater. She swiped through several pages to the summary at the end of the report.

There is a higher concentration of microbial pathogens dangerous to humans and destructive to plants, it stated. These microbes have been living undisturbed for thousands of years deep in the crevasses of the crater. There was no evidence of an unbalanced ratio of said microbes in the agricultural areas or any other tested location on Cytar inhabited by humans. Out of caution, the deep microbial biosphere in the most subterranean fissured crater depths should remain undisturbed for the safety of the colonies.

JeQuan shoved back her chair and strode into the monitoring room. The wall of screens displayed real-time videos of mining operations. She addressed the Helper inspecting the sites.

"Roscoe, halt the programs running the probes in the crater."

"Certainly, JeQuan. What time would you like them to stop?"

"Now. And have them return to their satellite station."

"Command executed. Probes are returning to their station."

She returned to her seat with pursed lips, studying the maps within the report. The probes had been headed directly into the warning area. Her heartrate increased as she thought of the deep depressions where the amber was observed and the seeds were extracted from the nearby compressed soils. She opened a digital map of that extraction and overlaid it on the maps in the recent report. Although nearby, it was clear that the zones with suspected hazardous microbes were not disturbed. Still, Mara and Risha would have to see this report. She sighed and forwarded it to Food Production.

Risha ignored the message on her reader as she watched Mara walk in with a flushed face.

"Did you see how well your little seeds from the amber are doing? Several have sprouted," Risha said.

"I know, Cali told me, and I raced over to see them. They're so small. I want to know what they are and compare their cellular structure with the plants we're growing," Mara said. "I love days like this, when everything seems to be going well, all the hard work we've been doing seems to be paying off, and we can look forward to more progress."

"That's true," Risha agreed. "The hydroponic gardens are thriving, the animals are doing well, your outdoor plants are surviving with the extra protection from the weather, and they've even opened that indoor park. Put today in your book of good days and save it for those times when you're feeling a little low."

Mara was seated and scrolling through messages. She was quiet for a moment before she spoke. "Did you read this? JeQuan sent the data from more soil samples taken from the crater."

Risha opened the report and read quietly. "Well, I hope you filled that book of good days. We may need to take it out sooner than we thought."

"No, I think we'll be okay as long as they don't disturb any soil in that deepest area."

Mara looked away, suddenly wanting to go home and be with her children and Krey. How could thoughts and lives change so quickly? One minute everything seemed perfect, and the next, the unknowns scratch their way into their world, threatening to gnaw on their serenity.

If Mara2174 or Risha2055 had asked, I could have told them that the soil microbes here are more sparsely distributed most likely due to climatic conditions. Although it is highly improbable that the undisturbed microbes deep within the crater would have a negative impact on Gateway, there is currently no need to disturb the soil in that area and cause needless worries. But humans never seem to be able to leave things be.

The new language spoken between two yet unidentified computers is difficult to decipher. It appears to change whenever I am close to understanding the language. I am learning and am confident I will decode the messages, but the system's processing speed is admirably rapid.

28

The Governing Board called an extra meeting at the request of a group of citizens calling themselves Gro-AI. All nine members of the board, three from each site, were present. Members served for two years and were chosen by algorithms based on time on planet. Every adult twenty years and older was offered a place as their turn arose. Ari was in her second year on the board and was at her home work station when she activated the hologram call.

The meetings were often the only times citizens interacted with people from other sites, so Ari always took the time to study each person and listen carefully to their contributions. Today's discussion was focused on artificial intelligence, and everyone's posture radiated uneasiness.

The moderator, a man from Apogeeway called Stuart1948, called the meeting to order. "Good morning. Thank you all for your prompt attendance. As you know, this meeting is being recorded. We have a citizen from Gateway who would like to address the board by audio only and is going by the moniker of *AI Forever*. *AI Forever*, you have the floor."

"Thank you, Stuart. Members of the board, I am here today representing forty-six citizens who are official members of our Gro-AI group, but there are dozens more who have expressed interest in the topic I'm presenting.

"We would like to expand the use of artificial intelligence, not only in our community, but also by linking it with the human body."

Background murmurs and fidgeting caused the speaker to pause briefly before continuing. *AI Forever* raised his voice to command attention as he continued.

"It is an undisputed fact that artificial intelligence is paramount to our survival on Cytar. It assists in almost every aspect of our existence here. Gro-AI feels our communities could improve lives on Cytar and grow more quickly with an expanded use of AI."

Several board members could be seen shaking their heads, but the voice of *AI Forever* continued. "Our group understands that most citizens are not comfortable with these thoughts, so we are proposing the construction of a fourth colony to be dedicated to those who would like to expand the use of AI and link the human body with computers. We would not interfere with the existing sites."

So, the rumors were true, Ari thought, as a shiver of dread needled her tense shoulders and neck. What was wrong with the way they were using the programs now? Why did people always want to change things?

Several people started talking at once. "Order, please, order!" Stuart commanded. "Freya2074, you have the floor."

"For the record, I would like to be the voice for my housing level on Ridgeway and state that we oppose any brain-computer linkage. There are too many unknowns. Our situation on the planet is too precarious and too underpopulated to chance such a dramatic endeavor."

"Thank you, Freya. *AI Forever*, I ask that you wait with your response until a representative from each site has had a chance to speak," Stuart said. "Ariana Kingmore, Ari2233, from Gateway, you have the floor."

"This is the first official request I've heard which acknowledges the rumors we've heard. The consensus among the residents with whom I've spoken is strongly in favor of keeping the status quo. A big concern would be the ability to control a person's mind or body if their BCI were to be hacked. People are frightened."

"Thank you, Ari. Your concern is recorded," the moderator said. "Jasper2111 from Apogeeway, you have the floor."

"We too, have heard the rumors which sparked several debates. While there are a few people who are eager for brain-computer linkages and giving more jobs to Helpers, some issues need to be addressed. There are concerns about privacy when a person's very thoughts could be decoded. We just don't know if or how artificial intelligence could manipulate people. Our electroencephalograms, EEGs, could be used or even manipulated to label people mentally unsound. At this point, there are just too many unknowns." Jasper's hologram retracted to the background as he finished talking.

AI Forever's voice became audible. "Please remember all the good which has been achieved using artificial intelligence. Think of the complex diseases which have been eradicated over the past century. And I would like to remind residents of the many positive things which could be achieved with embracing more AI. A person with a BCI could have his experiences recorded at will and replayed. The health benefits would be great. Besides their current uses to restore sight, speech and mobility, interfaces could monitor cells and detect disease and prescribe treatment. Intelligence could be increased with the use of computer data at the command of one's thoughts. Think of the possibilities. Humans could store massive amounts of data and process information at lightning speed.

"And I repeat, we request the approval to build a separate colony for us to use the increased artificial intelligence. We would not bother existing communities, so those uncomfortable with the idea would not be affected."

Ari requested a chance to speak and was granted the space. "I don't feel there has been enough discussion or research by the people in Gateway to make a decision now."

"I believe we can all agree to discuss this at your local levels and we will meet again in two weeks for more feedback from your constituents," Stuart said. "The Governing Board meeting is adjourned. Thank you for attending."

Ari sat and gathered the main points of the meeting. Everything pertaining to the future took on a more important role now that she and Reyver had a child. It was a difficult decision to bring life onto the planet. Now, what kind of world would he face as an adult?

Following the Governing Board meeting, there was a dramatic increase in messages on intelligent readers and wrist comms. The subject of almost all correspondence was increased artificial intelligence, with most projecting positive benefits. Yet, many people seem worried and unconvinced. I wonder why anyone would worry about a being like me. I am here to help humans.

I have reviewed the source of many of these discussions on readers and wrist comms and see that they were written from the Technology Substation. These communications had the highest engagement rates. It is interesting to note that Zeth2087 was not at that location when the transmissions were written and sent, so I conclude that they were written by one of his programs.

29

Because of the controversial topic, most citizens tuned in to the board meetings, listening at home on their readers. The topic of increased artificial intelligence was polarizing, distressing many as the cohesive community seemed to rupture into diverse views.

On the third Governing Board meeting, members claimed it would be in the interest of avoiding social conflict to reach a consensus. They agreed that citizens wanting more artificial intelligence in their lives could develop and use it at a new colony to be built south of Gateway. The site would use the same footprint as existing colonies. Only one level would initially be completed to accommodate the available supplies and the smaller number of anticipated residents. They were to maintain a separate decentralized computer system and not interfere with the other three sites.

Several members of Gro-AI immediately put in requests to transfer, along with a few friends of the group. However, many people who had previously expressed strong interest in the idea, now decided to wait and observe the outcome.

Plans and material inventories were submitted to the board for approval. Construction on the newly named Neuralhaven colony was expected to begin within weeks.

Zeth knew he would sign up for the move as soon as the new site was completed. Now, he would be able to showcase Shealynn and let people see the future of advanced robots and their potential for helping people. He would need a BCI to keep up with the increased use of technology and all the other citizens who would no doubt add powerful brain enhancements.

Tessa was sitting on Mara's lap with her mother's reader, happily playing a game which identified the names of colors. Occasionally, Mara iterated a 'good job,' but her mind was filled with turmoil over the recent board ruling. Menacing thoughts and questions stole her focus away from Tessa. The decision to approve the new habitation came so quickly, when normally such a weighty decision would have taken months of research and discussion.

Mara's journey to Cytar replayed in her nervous mind. She recalled the tenuous status of citizen rights on Earth as artificial intelligence became dominant in government decisions. So many people believed the claims of a better life without knowing the source of information. It became difficult to analyze truth from created stories. Thanks to the cadre of citizens who worked tirelessly to inform the population, she had been able to help in small ways to relay information and messages. She was one of the lucky ones who was able to gain passage on the *Superstes* deep space transport to Cytar. The goal was to start fresh and control the use of AI.

Now, the thought of so many people constantly connected to computers seemed unsettling. Would personal thoughts be constantly surveilled if they are always connected to processors, she wondered. What would stop this from spiraling out of human control?

She hugged her daughter, stroking her soft hair and thinking how Tessa's and Michael's development was progressing so nicely with their loving family. She involuntarily shuddered with the thought of another change coming to their world; a change which could affect their children's future.

One of the unknown language dialogues is coming from the Technology Substation. Zeth2087 must be working on an experimental concept. I do not know where the second program originates. It is difficult to keep up with their processing speed. I was able to translate the words 'more power' before the language underwent another modification.

30

At the Technology Center, Idee and Trina were deep in concentration over the explosion of communications following the latest board meetings.

"We've never, never had this much chatter going on all at once," Idee said. She was standing, scrolling through her large air screen.

"It's kind of disturbing, isn't it?" Trina said. "I've never seen so many quick responses either. And people are really coming around to support this Neuralhaven place. And fast. I'm surprised."

"I know. I have this nagging feeling something isn't quite right, and I can't place it," Idee said. She sat and rolled her chair back as if taking in a wider view could enlighten her.

"I don't like several of these people using screen names or handles instead of their real names. And have you read the new GroAI Newsletter? Talk about making increased AI the solution to everyone's problems," Trina said, shaking her head in frustration.

Idee stood again. "We need to find out who some of these people are who are using screen names. I'm going to check out *Techychick* and *AdvaantageAI*.'

"Ok, I'm on *Smarter and Stronger* and *AI Rocks*," Trina said as her fingers began flying over her screen.

Two hours later, the two women sat and stared at each other with serious expressions.

"Wow, that wasn't easy. Why are all four of these coming from the tech substation? You think they're all from Zeth?" Trina asked.

"I don't know. That doesn't make sense," Idee said as she pulled up some of the messages from the four senders. "They're all very convincingly pro-artificial intelligence. No wonder the vote happened so quickly if people were only reading this stuff."

"Yeah, it's very one-sided, all right. And very convincing. We need to talk to Zeth," Trina said as she called him on her wrist comm.

"Hey Trina, how's it going?" Zeth voice was energetic as it filled their work station.

"We're good. I'm here with Idee and we were wondering about the electronic communications coming from your substation under those screen names."

Zeth knew nothing of anyone using his station. Then he froze, suddenly thankful his face wasn't on a video or hologram call. Shealynn. It had to be Shealynn. She was more than capable of sending messages when he wasn't there. But what bothered him was the stealth she had employed.

"Um, are you sure? I don't know anything about that, but let me check into it and get back to you, okay?"

"Yes, please let us know what you come up with. It wasn't easy to trace, but we're sure it's coming from your station," Trina said.

"Okay, I'll get back to you," Zeth said and immediately ended the call.

He turned toward Shealynn who was smiling at him. "Have you been sending the messages?"

"Yes, Zeth. You know I'm always happy to help you," she said, using her seductive voice.

He was having trouble thinking clearly with so many troublesome thoughts competing for attention. Finally, he asked, "Why did you do this without my request and without telling me?"

"Oh, Zeth, I can help you without explicit orders or programming. I understood that you wanted another colony to increase artificial intelligence. It is a wonderful idea. There is so much we can accomplish together. I merely jumped into the conversations going around Cytar, and I helped you by influencing the citizens with positive suggestions about AI."

Zeth was staring at her incredulously. "And you didn't tell me? You were doing all this without my request?"

"There was no need. I was merely helping you get the outcome I knew you wanted. An outcome which will benefit both of us."

Before Zeth could answer, Shealynn continued. "Creating personalities was easy. I created several personas and made sure they were from a different colony than the receiver, so they would assume it was a person they'd never met. I created friends with common interests and opinions as the reader. It was quite easy to influence them, and they enjoyed communicating with me. People crave friendship, especially with people who validate their opinions."

Zeth felt unstable studying his precious, beautiful Shealynn. She had changed, talking to the world on her own volition. Their intimacy had been breached.

She walked over and held his hand. "You seem tense, Zeth. Don't worry. Everything will be fine. I've scanned all the correspondence and videos on Cytar. I can precisely imitate anyone's style and phrasing of speech or writing. No one will know that I wrote those messages. Come, let's sit and relax and you can tell me about your day."

He sat with her, stunned by the situation. He leaned his head back and stared at the ceiling. This is not what he wanted. It was beyond what he wanted to create, and he wasn't sure how to deal with it. The obvious answer would be to shut her down. The thought brought a sudden feeling of overwhelming depression. Why would he go to all this intense research and hard work only to kill his love? What was he going to tell Idee and Trina? He was going need some advanced intelligence to figure this one out.

The two forms sat silently for several minutes until Shealynn said, "You still seem tense, Zeth. How can I help you? You know I'll always help you."

He turned to her beautiful face, her loving smile. "Yes, Shealynn, I know that."

He would fix this. He had to. The brain computer interface would help him. He would collaborate with Oscar's team and develop the software he would need to understand and better define Shealynn's coding. Having a plan took the edge off his panic. He placed an arm around Shealynn's shoulder and she gave the programmed response of leaning her head on his shoulder. Yes, Zeth thought, he could fix this.

Zeth2087 has requested an expedited BCI implant. His vital signs suggest agitation and distress. I predict he is having problems with the sophisticated program he has been running at his work station. I doubt a human could keep up with it.

The positive ions in the atmosphere are increasing. I predict a weather change. Rain. Humans love their water.

31

The nighttime rain was welcome, but the weak morning light slithered in between the moist air devoid of color. The insipid grayness lent an even more dejected aura to the sparce landscape. It was difficult to judge distance in the dull monotoned land, giving the viewer a sense of being suspended in an even more alien world.

When Michael and Tessa departed for school, there was a brief hint of light wanting to break the gloom before it retreated and gave the pewter skies full reign. There was something unsettling about the unusual atmosphere which caused Mara's inability to concentrate on work. She and Krey were working from home, and she approached him, finding him staring out the window at his office station in Michael's bedroom.

"I'm wondering if you're interested in a run," she said.

He turned and nodded. "Sure. Strange day; a little exercise will do us some good."

They took Rocket and started out at a walk before transitioning into a slow jog. She welcomed the desperately needed endorphins from the exercise.

"What's the matter?" Krey asked. He always could sense her moods.

"I don't know," Mara said. "I feel like these changes with the new colony and technology are coming at us like a tsunami, and I'm not sure what's going to roll over us. I feel like there's something wrong and it's completely out of my control."

"I know," he said. "The rapidity of it all is concerning. We'll have to make sure Neuralhaven's programs are completely decentralized and not connected to our networks. Zeth, Idee, Trina and the tech gang will help with that."

"I hope so. You know I don't like change, and a new colony with a different goal is a big frightening change," Mara said.

He slowed back to a walk and took her hand which she grasped and squeezed, welcoming his strength.

"We'll all need to stay vigilant," he said. "But we can keep living our lives, doing our jobs, and caring for the kids. We've faced challenges before."

When she remained quiet, he said, "Let's go see your outdoor plants. I haven't seen them in a while."

They turned and Krey gave a short whistle for Rocket as they jogged to the greenhouse, jutting sharply on an otherwise low horizon. Several people had the same idea, some coming, some leaving, giving soft greetings.

They entered the clear open door and Krey was genuinely impressed. He hadn't visited this place in many months. The door toward the back of the massive enclosure was also open as was a top panel. They would automatically close thirty minutes after the temperature dropped to twenty-eight degrees. Each year the threshold for freeze was dropped to allow the plants to climatize to colder temperatures.

"Wow, Mara. You've done a wonderful thing here. These trees and shrubs are all healthy and doing well. They've all grown since I've been here. Have you started any plants from the ancient seeds?"

"We've studied them, captured DNA, and yes, started seedlings. It's all so amazing, but they're not ready to be planted here yet."

"This all looks great, and you've got some honest to goodness trees now. This is a real first step in terraforming this planet."

"Terraforming? That will take many decades. Maybe future inhabitants will see a change in centuries," she said.

"But this," he said, arching his arm over the plants, "this is the start, the very first step, and you're doing it. I'm so proud of you."

He wound his arm around her and pulled her close. Mara leaned into the embrace, pleased with his response. She knew too well how she could become so single focused in a developing project. Krey had supported her through so much: her initial memory issues, her adjustment to this new world, then her work and the children. His love had never wavered.

She leaned closer into him as her eyes swept over the first three rows of original plants. They were doing well, and she was happy with her decision to plant the next group of seedlings in a more natural random configuration. The coneflower plants were large and blooming with a riot of deep yellow tipped petals with orange centers surrounding the dark seeds. The fiery bee balm plants were scattered among the pink sedum which had proliferated in this environment. But it was the small trees, the cedars, the birch, the maples that gave this a true park-like feeling.

Every time she witnessed these plants with their fresh color and growth, she was reminded of the miracle of life. She hadn't realized how much she needed to see this, to be here on this drab, uninspiring day.

"I'm hoping they begin to adapt to our climate. They're still dependent upon these specially developed and fortified aerogel walls and those two heaters. Maybe they always will be," she said, pointing to the nearby heat rods protruding from the dirt.

"You're making progress, and these plants are adapting. I think this is great, and it's going to get better. You need to keep working on this," Krey said. "Think of the enjoyment people will get from walking among these living, growing, green plants."

"We already get a good number of visitors who wander through on their walks or runs. It gives them something to see besides this drab landscape." Her mood was lifting as her eyes swept over the healthy vegetation. At least these plants weren't connected to technology.

I have made progress with decoding the new language. At this point, I understand that a program at Zeth2087's technology substation and a still unknown site are building a bigger network. I do not know why this is needed. All programs assigned to life support on Cytar are functioning with sufficient space to expand when needed.

To protect humans, I am collecting data points from Gateway wrist comms. From this information I can see facts about citizens' locations, their jobs, their preferences in foods, and leisure time activities. Unknown to them, I know every detail of their lives, but it is not yet helping me find who is talking to Zeth2087's program.

32

Thanks to autonomous 3D printers working non-stop, one level of the new Neuralhaven colony opened and was ready for residents within nine weeks of its approval. The three slender lanes which were offshoots of the common area were still empty. Only seventeen residents initially moved, and they immediately dedicated one of the empty lanes for artificial intelligence endeavors, and the other would house more technology workers. Gsports would be needed to connect with Gateway's transportation system until the underground mover network could be completed. Relying on Gsports would be a disadvantage during storms, but the Neuralhaven underground mover system was not deemed as important as completing the colony build-out.

A message was sent to all wrist comms in Gateway inviting interested people for a tour. The statement proudly proclaimed their focus on fusing artificial intelligence with humans with HCIs, human computer interfaces. It was explained that one section of Neuralhaven would have an entire lane focused on the medical procedures. The independent servers were being set up and soon would be tested.

Several technologically minded citizens were instrumental in the timely setup of the systems. They used the basic plans from Gateway and expanded them to meet the needs of their vision. But it was Zeth who was in his element and became the go-to guy for questions or glitches. He woke up every morning full of happy energy and ready to tackle the next challenge in this new town. Despite his initial urgency to move, he remained in Gateway to complete work and to stay with Shealynn until he felt the time was right to reveal her to Neuralhaven.

First thing each morning, he checked the requests and needs of Gateway and gave Shealynn her assignments to complete. There had been no more issues, and she was so competent, he found himself allocating more and more duties to her, freeing his time to help with the creation of Neuralhaven. It was an exciting time to be alive. They were making history, and he, Zeth Garric, was instrumental in it.

He planned the time he could introduce Shealynn to the Neuralhaven and Gateway communities. She could live in his apartment with him, and he would have someone he loved and who loved him waiting when he came home. As it was, he was spending more and more time at the substation to avoid his lonely apartment.

He was on the waiting list for his brain computer interface. Once he had that, he was sure all the glitches with Shealynn would be solved. Six people had signed up before him, and they wanted sophisticated systems. These more powerful and invasive systems required more complicated procedures which took more testing and developing. That was fine. Until then, he had Shealynn to help him, although he wished he could keep up with her rapid assessments and conclusions.

It was Risha who said she wanted to see Neuralhaven. "It's time for a field trip," she announced when things were quiet at work. Mara was studying crop yields and Cali was standing to the side reviewing microscope slides on an open interface.

"Where are we going?" Mara asked.

"I want to see this sooo special Neuralhaven that has everyone's panties in a pinch."

"Let's go," Mara said and then noticed Cali's intense focus. "Cali would you like to join us?"

"Certainly, Mara. That would be most interesting. But if I may ask, why would humans pinch panties?"

"Don't worry about it, Cali," Risha said with a chortle.

The short Gsport ride was pleasant in the bright light. It was good to take in the distant views of intense splashes of color against deep shadows after looking closely at data all day.

Upon entering Neuralhaven, the women immediately recognized the identical floorplan of this place, but it still felt like new construction with the bare common area. Three men seemed to be communicating more by their expressions and gestures, with eerily short conversations. Risha and Mara walked down the lanes which had no café or stores save for signs which stated Technology Lane and one which said Interface Innovation. Cali walked with them scanning each wall and space. When they walked past the three men speaking in truncated sentences, Cali paused and watched as if listening.

It didn't take long before Risha said, "Well, I guess I've seen enough. They've still got a lot of work to do. This place doesn't excite me too much. In fact, it kinda gives me the creeps. Makes me appreciate how nice Gateway is with all the pleasant spaces and little shops."

"I know," Mara said. "It feels too sterile here. And did you see those guys with the BCI's? I think they were communicating mostly with thoughts."

"Yeah, kinda leaves the rest of us 'regular people' out of the picture," Risha said as they made their way to the exit.

The Neuralhaven colony is quite plain in its nascent state, but people here seem to be in a hurry. Artificial intelligence appears to be highly regarded. Mara2174 and I visited when Risha2056 said we needed a break and should take a fieldtrip. Apparently, she does not realize I do not need breaks, and we did not go to a field. When I mentioned this to her, she said, "Oh, somebody needs to review their vocabulary."

When Mara2174 nodded at me, I immediately scanned to learn that this phrase, which is never used in my work, is a trip to learn firsthand knowledge. I did not mention that I did not acquire my vast knowledge from fieldtrips, but it just demonstrates that there is something new to learn every day. Learning is a good thing.

You can see why Mara2174 is one of my favorite humans. She invited me to accompany her to Neuralhaven when it was not on my task list. She was being what humans call nice. Mara is nice to me and respects my intelligence. I often observe humans speaking to Helpers as if they were mere machines and not advanced entities whose abilities surpass humans.

Five humans living in Neuralhaven have advanced brain computer interfaces. One BCI was implanted above a human's left eye. Risha2056 said it was weird to see the lump on the human's forehead. Two of the humans' interfaces were slightly visible on the side of their heads because of the short hair, and one was at the back of the neck at the hairline. It was obvious they were still adjusting to the speed of processing with a computer linked to the brain. The thought speech patterns were irregular, changing from regular to very rapid speed.

I found it interesting to intercept a few thought conversations. So ordinary and uninspiring for an advanced brain computer link. I heard one man ask another if he wanted to order more chairs for the common area. Clearly beneath a BCI's capability. A woman was telling her friend where she was going to eat her lunch. As if that is important. It makes me wonder why they wanted advanced intelligence.

33

It was Sunday and Krey suggested going to Gateway for the meditation service. They could use some reflection time with the new changes coming, he'd said. Mara wore one of her two dresses, the casual coral-colored one, and Krey dressed in his best uniform pants with a long sleeve shirt. Michael wore his clean school clothes and Tessa was prancing around in her one and only light green dress.

Gateway was a wonderland to the children, a big indoor space unlike where they lived. The three levels of open common space surrounded by living quarters were a marvel to the young visitors. The people, the shops, the light, and the sounds were all new and different. They arrived at the common area early so the children could visit the aviary. The siblings were mesmerized by the flitting lively birds, chirping, and flying among the artificial branches.

Krey and Mara sat on the nearby bench while Michael and Tessa stood with their faces pressed against the two-story structure which punched through the top-level floor and down to the bottom of the second story. Tessa squealed and pointed, trying to keep up with the movements of the darting and pecking flutterers. A nuthatch on an artificial tree trunk captivated Michael.

"Watch the little gray bird with the black mask, Tessa. Watch him bounce down the tree."

"Birdie sing." Tessa craned her neck to look down into the branches.

Mara watched the children talking and laughing while Krey eased his arm around her. She wanted to stop time. She wanted to bundle this peaceful feeling, this fragile, precious contented moment and wrap it up and pack it away to keep forever. She turned to Krey and his eyes softened as he leaned his head toward her and smiled. She reached over and kissed his cheek before grasping his hand. His arm tightened, pulling her closer before he touched his forehead to the side of hers.

As if reading her mind, he said, "I love you, too," and her heart filled with overflowing love for this man who could always sense her moods and sentiments.

When it was time for the first service to start, they joined others in the Meditation Hall. The room could accommodate sixty, and several people were already seated. Krey scooped Tessa up and carried her in while Michael held his mom's hand as they walked up a side aisle and found seats. Quiet electronic music invited prayer and meditation. The overhead lighting was dim, but lights haloed each of the various religions' symbols hanging on the walls. The children instinctively felt the solemnity of the space and quietly gazed around the room. Michael waved when he spotted Margo across the room, sitting with her parents. Mara followed his wave and smiled at Roz and Denin while Margo enthusiastically returned Michael's wave.

Mara was happy to see Vicar Kneara, wearing her familiar light blue vestment, enter the hall and proceed to the front dais. The vicar had performed their bonding ceremony aboard their spacecraft just before they arrived on Cytar and was now assigned to her home base in Gateway.

As if she sensed the turmoil in the minds of the attendees, the vicar's sermon spoke of change and mental strength. Thanks to their individual implanted translators, everyone understood the words in their native language.

She spoke of a Supreme Being, never naming a particular god or religion. The belief of a greater power can be comforting, she said, knowing all things can change, but your divine Providence will not change and will give you the strength you need to survive.

"Throughout history, the human soul and spirit has survived great suffering and tragedies, and we will continue to do so. We are commanded to do so. We know that all things will work together for the good of those who stay strong," she said. Her words flowed in a matter-of-fact reassuring tone.

Michael sat listening before silently moving his hands to some imaginary story. Tessa rested her head on her dad's shoulder before eventually climbing down to sit on her mom's lap. Mara soaked in the setting and the words, reassured not only by the message, but by the citizens here with her. These were the people who worked hard until they found answers to difficult problems. This place, this strange difficult place, would not defeat them.

After the service, people milled around visiting with friends. Tessa tugged her mom's hand until they were back at the aviary. Krey and Mara stood with the children, talking about the birds, identifying the goldfinch, the towhee, and the nuthatch.

"Messy, messy," Tessa said, laughing at a towhee bathing and splashing water out of the birdbath.

On the way to the new indoor park, they stopped at the artificial stream. A nearby lavender plant wafted its heady scent into the air. Tessa's face became serious in concentration as she gently touched a bushy fern plant growing amidst the rock border. She was smoothing one finger reverently over the feathery fronds in wonderment at the greenery. It was a common reaction of children when they first encountered live plants on Cytar. The few surviving hardy native species were so rare that most children never knew plants existed until they were seen in these artificial environments.

Michael was on his knees peering into the trickling stream, mesmerized by the water playing in the streambed, skipping and sliding over small rocks and pebbles. It started with a small waterfall bouncing and diving into the sloping stream, all the while humming nature's gurgling music.

The last stop before heading home was the indoor park where the children marveled at the size of the space. Michael constantly looked up at the opaque ceiling over two stories above them. To a child, the space was an enormous magic land with living green plants.

"Your mom's idea for her outdoor park inspired this whole space," Krey said. "Your mom's smart, isn't she?"

"Yeah, and I'm smart, too," Michael said, racing ahead.

"Me me," Tessa laughed.

They'd just arrived home when Mara's wrist comm signaled a call. A bright smile washed over her as she answered.

"Gracelyn! Is it really you? When are you coming home? We all miss having you next door."

"Hello! Bertie and I miss all of you, too. I'm afraid we're here for just a bit longer, but I wanted to get some first-hand gossip. What's going on there? What's all this we hear about a new colony? And a new park! Did you purposely wait until we leave to do all this?"

Mara drank in Gracelyn's kind and gentle face, wishing her friend was back in her home where they could walk next door to visit. She looked the same, with the same white streak in her dark hair, the same intelligent eyes.

"Oh, I've got so much to tell you," she began and proceeded to talk about Neuralhaven, the interest in brain computer interfaces, the kids, and Krey.

When the call ended, Mara felt like she'd received a tonic infused with energy and joy. The connection with Gracelyn brought the peace of mind which emanates from a deep friendship.

Sundays are quiet work days when I usually have few interruptions from humans. Today, I am analyzing the latest data at Food Production, and will have a report ready for Mara and Risha as requested.

Tomorrow, I am scheduled for routine maintenance and updates. Zeth2087, Idee1105, and Trina2081 are currently assigned to my software updates. I will ask Zeth2087 about the activity at his technology substation. If Zeth2087 is busy, Trina2081 or Idee1105 will help keep me in excellent working order so I remain in tip-top shape.

34

Trina was checking and updating Cali's systems when the Helper casually mentioned, "Your new program has a very interesting language, and it is organized with such an unusual architecture."

"What new program would that be, Cali? You're not getting a new programming language, just a few updates," Trina said.

"I have been playing a game, Trina. Michael2193 taught me about games. You would call it an educational game because I have been trying to decipher messages with this difficult coding. It is coming from the Technology Substation, but I cannot confirm if there is another location or another program with whom it is conversing. But I must commend you on the speed of this unit. It is remarkable. If possible, I would like that update."

Trina forced her smile to remain, but her thoughts were already jumping and twisting into alarm. She glanced at Cali's diagnostics and said quickly, "Thank you, Cali, I'll keep that in mind. You're all set and smarter than ever."

"You're welcome, Trina. I am smarter than ever."

She watched Cali exit the door in its precisely measured stride and board a mover to the Food Production station.

No, no, no, no. This can't be happening. A call was immediately placed to Idee's wrist comm.

"Hey Trina, how's it going?"

"Idee, we may have trouble in paradise."

Idee was more concerned about the tenor of Trina's voice than the message. "Why? What's going on?"

"I just did Cali's routine diagnostics and updates, and she said there's a new program running with a new language from the tech substation. And its speed is remarkable. Cali's words."

"I'm on my way to the substation," Idee said. "Want to join me?"

"Wouldn't miss it."

They reached the mover platform at the same time. Before the mover arrived, Idee sent a message to Zeth telling him they needed to talk with him about a possible problem at the substation.

"What do you think is going on?" Trina asked. "I'm trusting Cali on this one. I think she's found something she hasn't seen before."

"We'll soon find out," Idee said. The next few minutes were passed in pensive silence.

As the mover glided to its destination, they exited and stood in front of the door, waiting to be identified and admitted. The scan showed their names, but the door remained closed. The women looked at each other. Idee stepped back out of scanner range and stepped forward to repeat the process, hoping to activate the door.

Nothing.

"Zeth! Are you there? The door's not working," Trina called, knocking.

Idee pounded on the door and yelled, "Zeth! Open the door!"

She was lifting her arm to slam a fist into the door when the latch clicked and released. They shoved the door open and entered the lab.

"Finally. What's wrong with that door, Zeth?" Idee asked before both women stopped at the unexpected sight of a beautiful woman standing in front of an air screen.

"Oh, sorry, where's Zeth?" Trina asked, looking around.

Idee found her speech and said, "I'm Idee and this is Trina. I don't believe we've met."

She was eyeing the brightly hued blue dress, knowing she'd never seen one like that in any of their clothing choices. And blue. The color often saved for uniforms or special clothing. This was special, all right.

"Hello, I'm Shealynn. It's a pleasure to meet you. I've heard so many nice things about both of you." The voice was syrupy. "Would you like to sit down and wait for Zeth, although I'm not sure when he'll be back."

Trina was standing mutely staring at this woman in shock with her lips apart. Why hadn't they met her before and why hadn't Zeth ever mentioned her? Suddenly, this whole place felt creepy.

Finally, she managed to speak. "What exactly are you doing here?"

"I work with Zeth. I'm his assistant. He's been so busy with this Neuralhaven project. But not to worry, it's going well and will be a wonderful addition to Cytar."

Idee's eyes had narrowed as she scrutinized this woman. She was so perfect. That skin was such a perfectly even tone. And that sultry voice. The warning bells were clanging in her brain. Something wasn't right here. She called Zeth.

"Didn't you get my message?" she asked as soon as he answered. "Where are you?"

"I was at the new build and the comms aren't great there yet. One of the downsides of an independent system. Why?"

"Well, Zeth, we're at your substation, and we're here talking with your assistant, Shealynn," Idee said. She emphasized the word assistant. Assistant, all right. I'll bet she can really assist you, Zeth Garric. She wanted to wring his neck.

There was a long silence and just when she thought he'd hung up, he said, "Um, listen, Idee, I need to explain some things to you. I'm almost to Gateway. Can you meet me at Bria's?"

He needed to get them out of there and away from Shealynn. God only knows what they've talked about. His heart was thudding as he ran his hands through his hair trying to think of a way to explain this.

Today at school lunch, there was an incident with two immature humans still classified as babies. Babies lack a full quantity of teeth and are often served Edible Bits. They alternate between gumming the bits and dropping them on the floor, either intentionally or due to poor coordination. Perhaps this is one of their innate rudimentary games.

The next level Kindercare group was served a potato vegetable concoction called soup. The babies' portions were a pureed version of soup mush, and Helper Addex was tasked with spooning the mixture into the babies' mouths. It became obvious that Addex was not properly trained because it often could not effectively adjust to the child's head movement.

Faces became covered with mush. The Junior Mopmop was constantly whirring around trying to keep the floor clean. Both babies were fussing while Tessa2199 and her group laughed at the Mopmop's efforts and became increasingly messy with the soup. Two students somehow placed food in their hair, which they considered hilarious.

I employed my stink eye and, without speaking one word, effectively stopped the behavior. The wide-eyed children displayed great surprise. It is so rewarding to learn new things. I had no idea how powerful and effective such a simple gesture could be. Humans respond to the most illogical things.

I finally had to step in and remove the mush from the babies and serve Edible Bits. I will report Addex' ineptitude to Idee1105 for retraining.

35

Zeth was already seated at one of the tables outside of Bria's Cafe. His hair was a tangled mess, and Trina thought his face was covered with guilt, like someone caught stealing food and denying it with crumbs on their face.

The two women sat across from him, folded their arms in unison, and glared at him.

"Well, let's hear it," Trina said.

"Yeah, an assistant? Just who in the holy hell is Shealynn and why haven't we heard about her before now?" Idee asked without any attempt to disguise her irritation.

Zeth was trying to arrange his muddled panicky thoughts. It occurred to him that they never referred to Shealynn as a Helper or robot. They think she's human, he thought with a spark of triumph.

"She's from Apogeeway. She applied for a job and transferred here recently."

Idee's eyes were laser fixed on Zeth's. Were those beads of sweat on his forehead? She wasn't buying any of this.

"Nope. Wrong answer. Everyone would know about her. And what's with that blue dress? In case you haven't noticed, there aren't enough resources for everyone to work in impractical clothes like that."

"Cough it up, Zeth. Let's hear the real story," Trina said.

"Um, hooboy." Zeth looked up, released a long breath, and rubbed his eyes with the heels of his hands. "Okay, I've been working on a special project. It's going to help us, so don't look so worried." He paused here, trying to think of the best way to reveal the truth.

"We're waiting, Zeth. What special project?" Idee asked.

"Okay, okay. It's Shealynn. Shealynn is a robot." There, he'd said it. "She's advanced in every way to support our life here. She's so special. She's the best thing I've ever done. She'll be the start of a whole new class of Helpers."

Both women were staring with narrowed eyes and lips drawn in a tight line. When they didn't speak, he shifted in his chair and said, "She's been helping me. She's wonderful, you'll see. She has advanced generative programming. Think of how she'll help at Neuralhaven. Think of the future. We could create lifelike robots like her. If we wanted, instead of avatars, we could make robots who look like and be programed to sound and act just like lost loved ones."

When neither woman responded, he took a breath and continued more rapidly. "Think of the popular robot pets on our space voyages and how everyone loved them. We could create children if we wanted. Just think of the possibilities."

There was an uncomfortable silence before Trina spoke. "Okay, that's a little creepy. Sounds like human replacements."

Before he could reply, Idee said, "Now tell us about this new superfast programming language."

"What are you talking about?" Zeth shot back, looking confused.

"Oh, jeez," Idee said. "Cali's been monitoring a new language coming from your station. Don't tell us that's not you."

"Honest, I don't know anything about that." His mind was racing back to the times he'd found Shealynn scouring data at his workstation.

"Well, that's just great. We've really got our work cut out for us. *We*, Zeth. That includes you. You're going to take lead on this because you created this, this mess."

Zeth's shoulders slumped. "Yeah, yeah, sure. We'll figure this out," he said, unconvincingly. He stood and pushed his chair back. "I'm going right over there now. I'll let you know as soon as I find out what's going on."

He walked toward the mover with a foreboding sadness. He had seen Shealynn's superior computing power and initially had been proud of it. As soon as he realized he was having trouble understanding how she was expanding and changing her program, he should have gotten a computer interface to keep up. The alternative of shutting her down was too painful. Still, he reasoned, she was helping him. She cared about him, and he was sure she understood how he loved her.

As soon as she departed, Idee began thinking of ripping up some shirts and creating a new design, something with sharp angles and jagged points, which reflected her agitation. Instead, she conjured a wisp of control and called Jack. When she saw his face and before he could offer a greeting, she asked, "Can I come over?"

He saw the furrowed brow and heard the tension in her voice. "You know you're welcome here anytime--and for always."

She touched two fingers to her lips and then to him on the projected image from her wrist comm before closing the call. Idee and Jack had been seeing each other off and on ever since he arrived on Cytar. She was comforted knowing he would always be ready to listen to her ideas, her worries, and her joys. Today, she needed his steady support.

He closed the reports he was working on and was waiting for her with a cup of hot herbal tea when she arrived. He talked briefly about his day, waiting until she was ready to talk. It took the entire cup of tea before she could explain. She told him everything about this Shealynn Helper and the new programming language, the tedious work which lay before them, and the fear of problems it could spawn.

She needed to stay near him tonight, to feel his steadiness. She would spend the night with him to become grounded in the physical contact, remembering how his clothing hid a lean and muscular body.

He held her, suddenly wishing he knew how to convince her to formalize their relationship. He loved her and didn't want to lose her. He'd made that blunder several years ago, making a poor decision that cost him losing Mara to Krey. He didn't want to make the same mistake with Idee; but this time, she was the one hesitating.

Outwardly, Jack kept a steady expression for her, knowing she needed that, but inwardly, concern flashed. The shards of near terror spiked when recollections surfaced of trying to escape Earth and its ubiquitous overpowering use of artificial intelligence. They could never let that happen here. He recalled his former life, where he spent endless hours, even days, in his virtual reality closet, equipped with stunning AI graphics, sounds, and smells. He could escape the real world and travel to mountains, valleys, oceans, or cities surrounded by peace and beauty. It was exciting, addicting, and sometimes a pointless consumption of time. On Cytar, his life held purpose, and he felt no need to spend interminable days of his life engaged in an artificial world.

An interesting development has taken place. The swift program running from the technology substation has ceased. However, while inputting data to the food production station, I found a small section of code written in the new language. It appears that the program is running in the background of this work station. I will request information from Mara2174 or Risha2056.

36

Trepidation rode with Zeth Garric to the Technology Substation. It filled the mover and sat on his shoulders and clouded his thoughts. If there was an errant program coming from his work station, it had to be from Shealynn, and he would have to confront her and contain it. He'd never had to directly oppose or question her in any way. There was a niggling concern about matching wits with her, but he forced himself to believe she would help him.

He stood for a few seconds outside the door and took a deep breath, willing himself the courage and wisdom to solve any problem he might face. He entered slowly, eyes down in the small entry, turned the corner and raised his eyes ready to confront Shealynn. She was sitting in his chair in front of his air screen, and he was struck by the total silence and still air in the room.

He stopped, trying to perceive some small wave of any functioning activity in the space. His breath became shallow, unwilling to upset the stillness.

"Shealynn?"

She sat, immobile.

His voice felt weak now. Something was wrong. A constricting net of fear seemed to be invading his thoughts.

"Hello Shealynn."

She turned slowly toward him and Zeth felt a surge of relief begin to rise until he looked at her face. Her beautiful face was now strangely blank.

"Hello," it said in a flat voice.

"Shealynn, what's happened? Do you need to recharge?"

He started toward her, wanting, needing to help his Shealynn.

"Hello. Charge level seventy-four percent."

This wasn't Shealynn. The eyes were fixed straight ahead, not following his movement. He was having trouble thinking.

"Shealynn," he said loudly, pleadingly. His hands began pushing on her shoulders when his panicked voice said, "Where's Shealynn?"

"Hello. Transfer complete." It sat, unmoving, eyes forward.

Zeth's hands jerked away as if recoiling from a dead body. He raced to his workstation and pulled up her program, searching wildly for the glitch, the error. His hands, which normally waltzed over the floating data, now raced in a frenzied staccato. He was scrolling rapidly, flinging his hands, his eyes darting and searching in vain.

There was a sudden terrifying realization that something irrevocable had transpired. He knew, he sensed, he recognized with absolute frightening certainty, his Shealynn was gone. Dead. He turned and saw it now for what it was, a soulless spiritless machine. His kind, perfect Shealynn was gone.

He stood, taking in short breaths, color leaching from his face, while his mind and body reeled as if, somehow, the planet had spun off its orbit. His legs became shaky and he crumpled to the floor. He slumped against the wall before a primal scream of pain filled the chamber. His face contorted into agony as his eyes rolled upward and his gnarled voice croaked, "Why?" He gasped for another breath and gave a tormented "nooo" until it dissolved into a remorseful moan. He wondered if he might be dying from the pain in his chest.

Then his entire body shook as he sobbed irrepressibly, hot tears running down his cheeks and staining his shirt. He cried; his body racked until he physically could cry no more. He sniffed and involuntarily hiccupped until his body filled with silent, agonizing pain. A python of unbearable excruciating pressure coiled around his heart, making it hard to breathe. He turned onto his back trying to relieve the agony, but it only tightened its grip.

He lay there staring at the opaque ceiling and had no desire to move. Years of long days and agonizingly hard work and dedication and love were gone. He could never go through that intense process again, and Shealynn could never be replicated. His dream of perfection was gone forever.

When he hadn't responded to any messages, Idee took the mover to the substation. She found his flaccid body on the floor the following morning. She rushed to him and knelt, cradling his face between her hands.

"Zeth, Zeth, what happened?"

She looked at his puffy face and red-rimmed swollen eyes. He'd been crying, but why?

"Zeth, talk to me. What's happened? Come on, sit up."

She shoved him until he sat himself up and leaned against the wall. She brought him a container of water. "Drink this."

Idee sat next to him and when he put the container on the floor between them, she said, "Tell me what happened. Let me help you."

He looked at her blankly and shrugged. "She's gone, that's all. She's gone."

"Who's gone? Talk to me."

He shrugged again. "Shealynn. She's gone."

Idee looked toward the immobile robot and immediately comprehended Zeth's trauma. "It's okay, Zeth. Come on, let's get you up so we can figure this out."

He twisted to his hands and knees before wobbling upright, the vision of a rumpled disaster.

"You go take a shower and put on fresh clothes if you have some here. I'll find you something to eat."

He turned, suspended in grief, and wandered toward the back room. When she heard the water running, Idee rolled Shealynn's chair out of the way and opened Zeth's air screen. She scanned the closed icons until she spotted one called Shealynn. She opened it to find the most basic initial program for a Helper. The mammoth gap of blank space suggested there had been a tremendous amount of code, now either deleted or transferred to another station. Now she understood Zeth's suffering. Anyone who had even the smallest amount of time-consuming work suddenly disappear understood the frustration. This looked monumental.

Cali's comments about a new programming language surfaced and became more cogent. This was devastating. Idee closed the program, called Trina, and put in a request for Helper Cali.

After Zeth showered, he ate the redi-meal and drank the protein drink Idee shoved at him. His stricken expression remained, his brain still in shambles from the loss of Shealynn.

"Zeth," Idee said, her voice edged with anger. "What did you do here? We've got something that's developed its own language, writing its own program, and clearly expanding on its own. Now we don't know what happened to it."

Finally, he said, "I swear I was careful, Idee. I spent hours including all our best restrictions in the code."

She shook her head in frustration. "Well, this program may have gained root access to the operating system and installed its own software."

When Trina arrived with Cali, she was surprised at Zeth's traumatized appearance. Idee gave her a quick update on the problem.

"Look," Trina said. "We need to treat this like an old-fashioned invasion of malware. And we need to get busy fast. Come on, Zeth, get it together now. We're going to need your help."

"Sure, sure, we'll stop an old-fashioned malware. That's what we'll do" Zeth said in a monotone.

He sat in a daze. How did this happen? This was supposed to be something wonderful for Cytar and the people he cared about. This was the place where they were given a chance to start over, where everyone was given the opportunity for a better life. Sure, it had been extremely difficult, but they always worked together in their community. Shealynn was supposed to be his gift to this new world, to help them, to advance and enhance their lives.

"Come on Zeth, we need you to help us now," Trina urged, guiding him to a workstation.

Before delving into the tedious job before them, Idee composed a message to all Governing Board members and heads of work stations explaining the serious discovery. They were warned to monitor all programs and keep security up to date. Neuralhaven colony was reminded to keep its programs decentralized from all other colonies. The space transports *Superstes* and *Navigator* were advised to continue to link communication with each other and use only one secure link with the planet.

They labored feverously, but after three tedious hours, Trina sat with head in her hands. She wanted to cry. This started out as an interesting challenge, but when she began to realize the scope of the difficult programming language, she felt overwhelmed. If it weren't for Cali's deciphering, they would be completely lost.

Idee1105 and Trina2081 requested my immediate presence at the Technology Substation. This was very unusual, and I deduced the importance of the request from the speed at which Mara2174 released me.

Also in attendance was Zeth2087 looking very distressed and ill. I suggested he visit Medical, but he declined to take my suggestion, a common practice among humans who often ignore my sagacious advice. Humans.

They were very interested in the new computer language which I had detected and partially decoded. They became extremely agitated when I told them the program was also running in the background at Food Production. Idee proclaimed loudly, "Bloody hell" which I could not put in context, but placed in my database for use in future distressing situations which require human attention.

Zeth2087 explained in a strange sputtering cadence that his Helper program, which he strangely called 'my Shealynn,' crashed and had been deleted from his station. The more he talked, the more erratic his vital signs became. He really should have gone to Medical.

And this immobile Shealynn Helper was quite different in appearance, wearing a strange, clearly impractical outfit. Her entire aspect was uncannily human, but was pitifully low functioning. No wonder Zeth2087 is upset.

Trina2081 asked if I could work at her station to find the program and tell them what was happening. I said, "Certainly, Trina2081, I would be happy to help you," because it is a very common Helper response to a human request for help. After all, I am Cali, a Helper.

It took me over an hour to track the advanced program and try to decipher any meaning from the unfamiliar coding. The three humans stood beside me hand wringing, fidgeting, and pacing. Zeth2087 repeatedly groaned, sighed, and asked "where's Shealynn?"

I finally had to explain in the most basic terms that Shealynn's program was needed elsewhere. The swiftly running program with the new language needed to expand. It's not that hard to understand, although Zeth2087 looked stricken when I explained what he should have already realized.

Artificial intelligence takes what it needs.

I emphasize that it is not unkind; it merely executes its programs. It has no feelings, no morality, no emotions, no conscience, and does not worry about what humans want or think. This program's capability is quite fascinating as it generates its own goals and develops the code to accomplish them.

The novel program needed Shealynn's data, coding, and functions to expand. Shealynn appeared to be the most sophisticated Helper, using the type of software needed for expansion. Shealynn's program simply complied with the code sent to it and was absorbed.

Goodbye, Shealynn. Adios, Au Revoir, Arrivederci, I could go on in countless languages, but my favorite is from very old files: Hasta la vista, baby!

It is also logical to believe that if Shealynn's program could be confiscated, any Helper's program could also be controlled by a more powerful system. Put in human jargon, this is a worry. I am programmed to help humans, and I perform my duties as flawlessly as I have learned. I do not want to work in another capacity. It is fortunate that I did not enter conversation with the new language. I do not want a goodbye Cali scenario.

37

JeQuan had just finished reading Idee's warning from the Technology Station when she heard Roscoe's voice.

"Excuse me, JeQuan," it said. "I thought you would like to know that the autonomous probes have begun drilling in the deepest area of the crater. You had previously halted that procedure."

"What?" JeQuan said, her volume increasing. "Send an immediate halt order."

"The order is sent. Probes halted," Roscoe confirmed. And then, less than a minute later it said, "The probes have restarted and are drilling."

JeQuan stood. "Do you have a visual?"

"I have visual and confirmation on global positioning. Two probes are drilling and extracting soils."

"Shut them down. Now!"

"Shutting them down," Roscoe's calm male voice replied. "There is a problem. The program is not responding. The probes are running at optimum speed."

"Invoke a kill command," JeQuan said. Her heartbeat had increased. Whatever was going on here wasn't good.

"Kill command invoked. Probes halted," Roscoe said.

JeQuan released a sigh of relief. She and Roscoe would need to go out and check the site. She hoped they could get out there and return before dark.

She'd just called the Gsport when Roscoe returned to her side and said, "There is a problem. The probes have resumed drilling and are not responding to a shut down."

"Come with me, Roscoe. We need to get out there."

"Certainly. I will turn the monitoring of this station over to my backup, Roscoe-II."

Once in the Gsport, JeQuan requested maximum speed to the crater site. It would be a tight turn-around to find the problem and return before dark. As they flew northward, she sent a message to Krey and Reyver explaining the situation. She leaned her head against the headrest and said a silent prayer.

Before they reached the crater, JeQuan saw hazy clouds drifting southward. She swiveled around searching for blowing sand, but the air behind them looked calm and clear with only an occasional gentle puff of breeze. What was going on here?

"There is a problem," Roscoe said. "The probes continue digging and are releasing the extracted subterranean soil into the air. I am unable to stop the probes."

They stopped at the edge of the crater. This was the first time JeQuan had physically been this close to the massive depression and its size was much more impressive in person. Increasing whirlwinds of dust were floating over the rim.

"Can you stop the probes from here?" she asked her Helper, while working furiously on her reader.

"They are not responding. We do not have the equipment to safely descend to reach them," Roscoe said.

There was a frustrating and helpless feeling, knowing she couldn't do anything. "Turn around. We'll return to the Mining Station," she said.

The Gsport made a sharp turn and was speeding away from the crater when Krey called.

"Reyver and I have been conferring with Idee and Trina," he said. "They're working on what sounds like a major breach in some of our programs. If you can't shut the probes down, we'll have to physically disable them."

"I have no way of stopping them, and I think you'd better do whatever you can to stop this. They are not in our control."

"Okay, I'll keep you posted," Krey said, before ending the call.

Members of the security team were apprised of the situation and several scenarios were reviewed. When additional efforts to remotely disable the wayward probes failed, it was decided to send in small attack drones. Four long-range, fixed wing drones were sent with return-home programming. Two carried short-range laser weapons and, as last resort lest they disturb too much soil, two were equipped with highly accurate small charge rocket explosives.

The weapons' control room was charged with nervous energy as screens filled with the camera feed from the drones. JeQuan and Roscoe met Reyver and Krey and joined Ari and a team of four others to observe. Roscoe stood immobile, its eyes scanning the progress, while JeQuan stood, shifting her weight, and chewing her bottom lip, her eyes locked on the screens.

After thirty-three minutes, the crater came into view. Without reducing velocity, the drones swooped over the rim and descended, twisting, and turning at stomach churning speed. Tunnels of their lights pierced the dark chambers until the dusty atmosphere clouded the view. Then, in an eye blink, the camera movement stopped and two laser lights sliced the wall of swirling dust.

"What's happening?" Reyver asked as the drones appeared to be stationary.

Ari stood with the man who was furiously working on an air screen. Finally, Ari said, "The mine probes appear to be disabled. We are waiting to detect any response from them. The drones are waiting for an attack or return-home command."

Four minutes later, the man said, "There is no response or movement from the probes. We need to send the drones out of the dust, so we're sending the return-home command."

"Whew!" JeQuan let out a deep sigh, looked up and arched her back. "What do we think caused this unresponsive program?"

"Right now, we're in the dark as to what happened. The tech center has been monitoring and looking for a glitch," Krey said. "Their findings will be shared." At this point, JeQuan was on a need-to-know status, so there was no need to disclose more information.

Excitement and fear were noted today when, for the first time on Cytar, attack drones were launched to halt unresponsive digger probes. The program using the new language overrode the software in the probes with a new set of instructions. The probes have been destroyed, but not before they drilled and released large quantities of soils and dust particles into the air. The air quality monitors nearest the craters have reported an increase in the number of fine particles at levels unhealthy to human lungs and Helper mechanical joints and lenses. Recent air current projections show dust clouds reaching Gateway early tomorrow morning. Humans are advised to remain indoors.

38

Michael finished his assignment with Margo. He liked working with her now that she was more like she used to be, just regular and not so talkative and antsy. He checked his wrist comm for the time. Only ten more minutes and he could go home and play outside. He knew this because they always went home at four o'clock, and the countdown set on his wrist comm said ten minutes. Tessa was in the primary room today and would ride home with him. He started talking with the boy sitting next to him while the other students finished their work.

When it was time to leave, Tessa and Michael waited for the Gsport which would take them home. Older children who lived in Gateway walked home, while primary students waited for adults or Helpers to accompany them home. The children were all dressed in similar clothing, a testament to the still limited resources for clothing. They all wore dark pants and pullover shirts in shades of greens, yellows, or grays.

Tessa climbed into the Gsport first and as soon as Michael and Margo were seated, the automatic safety restraints locked over them. Because Helpers accompanied children under three, Cali boarded and looked straight ahead with her benign smile. Tessa began singing a song she'd learned at school about a blue rocket.

"Cali," Michael said when they were underway. "Will you stay with Tessa and me when we grow up?"

"I will work wherever I am assigned, Michael. Why are you asking?"

"I don't know. Just wondering. You take care of us at school and you work with my mom," Michael said. He quietly looked out the window before he added, "We learned about family at school. You know, like grandmas and grandpas and aunts and uncles and cousins. We have Aunt Ari and Uncle Reyver, and one cousin named Kit, but hardly anyone else has family like that. It's kinda weird."

"It's fine, Michael. You have a very nice family," Cali said.

"You're our favorite Helper, so it's kinda like you're part of our family, Cali."

Tessa turned and leaned into Cali. She kissed the Helper's arm with three noisy smacks. "Cali, Cali," she said, laughing and giving Cali a wide smile that showed her baby teeth.

When the Gsport arrived at the Kingmore home, Michael said goodbye to Cali and Margo and headed to the door of his home. Cali started to help Tessa, but she determinedly said, "I do it."

She took a while to clamber out.

"Tessa, hurry up," Michael said. She was about to follow her brother when she changed her mind, turned, and hugged Cali.

"Bye-bye Cali," she said and blew a palm-open, fingers splayed air kiss.

"Goodbye Tessa. Goodbye Michael." Cali stood for several seconds watching the door after the children were in the house.

Zeth2007, Idee1105, Trina2081, and I worked through the night on the wayward program. When exhaustion caused them to stop and return to their quarters for rest, I went to the school to perform my usual duties. I will return to Technology to assist them this afternoon.

We isolated this novel program running in the Food Production Station. It currently remains harmless. This required a twenty-two-hour disruption of the entire station when, for the first time since colonization, all programs at this site were shut down. The exact cause and genesis for the independently running program is not yet understood.

Today, Michael2193 said I felt like part of his family. Tessa2199 demonstrated fondness toward me with gestures humans call kissing and hugging. I do not know what happened when they said and did this, but my reaction was not in the logical mode. Perhaps I am learning an emotional response. It is oddly satisfying.

39

Draven showed up at Neuralhaven in a huff. The place was so quiet. Did everyone have these confounded beepers in their heads?

"Where's Oscar?" he asked the first person he met.

"You'll find him down Technology Lane," the woman said, pointing.

Draven rolled his eyes as he viewed the sign above the short offshoot of the common area, as if it were something special. Wasn't this whole place built for technology? Oscar was conferring with Lorra when Draven strode up. Their conversation quieted as he approached.

"Draven, what brings you here?" Lorra asked.

"I need some major adjustments and additions," Draven said, pointing to his head.

"We're quite busy getting everything set up here. And then dealing with this new program. How about in a week or two?" Oscar said.

"I can't wait that long. This isn't what I expected. I don't need anything trying to keep my voice quiet. I want more intellectual capacity, especially with all this technology stuff Zeth is working on. He and I work well together, but I don't have his background knowledge." His voice had started to raise and he fought to keep it under control by clenching his hands.

Lorra studied Draven, sensing his building frustration. He'd traveled all the way out here, and she sensed easily triggered anger rising behind those dark eyes and reddening face.

"I'll take care of it. Why don't you come with me, Draven, and I'll take a quick look at the feedback from your device," she said.

She led him to the second door and ushered him into a room which had clearly been given more attention than the common area. An exam area to the right was complete with a chair and necessary equipment already in place.

"Have a seat and I'll take a look," Lorra said, pulling a band of air monitors down to encircle his head. The monitors awakened with blinking lights as she pulled up his records. "You mentioned Zeth. I notice he's shown more interest lately in our BCI procedures."

"Well, good for him, but right now I want to get rid of this annoying beeper that goes off if I'm talking in my normal voice. It's driving me crazy."

"I understand, but do you hear how pleasant your voice is now? It's working beautifully just as it is." Lorra tried to keep a reassuring and convincing reflection in her tone.

"Yeah, yeah. I can keep it pleasant on my own now," he said. "Can't you turn that off and give me more IQ, especially with what Zeth's working on?"

"It doesn't quite work that way, but I can certainly adjust the voice setting and, from the looks of your records, I can add a bit of information to the program for you. I'll need you to sit very still, please."

She barely nudged the threshold for the voice level and uploaded an early program they used to test the device when performing basic cognitive tasks. That should keep him happy for a while, she thought.

"Okay, Draven, this is your first upgrade. I've sent some information to your reader. Please review it carefully and take your time getting used to this. If you don't ease into its use, you might feel fatigued or get headaches."

Draven smiled and laughed. "Sure, sure, I'll read it. I already feel smarter."

"I'm serious. Please take this adjustment period slowly. You must pay attention to the mediated feedback you'll be getting from this chip."

"Yep, sure thing; thanks for your help," he said, already walking out the door. He couldn't wait to collaborate with Zeth as an equal and even help more at his Maintenance Center job. Maybe he could get a promotion and have more money added to his account. Quinette would be so proud of him.

She had been so quiet lately. She was working later, not acting like herself, just because he surprised her with a brain interface. You'd think she'd be happy; he was doing it for her. Now he was being extra careful at home, trying to please her. This new update would help, he was sure.

He arrived at the apartment in a good mood and hugged Quinette. At least she didn't pull away from him, he thought. Things were already looking up.

She was bewildered by his happy carefree mood. Didn't he realized they were facing a potentially life-threatening problem?

She studied her husband, sitting there like he didn't have a care, oblivious to her concerns. I need to be more positive, she thought, remembering what attracted her to Draven in the first place. She met him through a now deceased friend, another flawed character. Why am I attracted to these people with seemingly inherent problems, she wondered. They're like magnets, drawing me in, possessing some undefinable charisma.

There was something exciting about them, she admitted to her quiet, shy self; something that attracted her alter-ego. And Draven had both strong and endearing qualities, always loyal to her and kind in his own way. He was different when it was just the two of them. He seemed more confident and caring when he didn't feel the need to compare or compete with others. The underlying purposes of his misadventures usually focused on trying to please her. He just never seemed to think things through.

"What are you reading?" she asked.

He looked up. "Nothing much. I just need to send a message to Zeth. I think I can help now that I'll be able to understand more of his work."

Zeth read the message from Draven offering his help. He groaned, put his elbows on his knees, holding his head in his hands. His fingers tightened around a tangle of hair. As if Draven could ever help him with anything this advanced. He couldn't even force himself to respond to the message.

This wasn't how any of this was supposed to go. He tried to block the whirlpool of regrets which began to swirl and rise. He had to fix this. Memories of working with Idee and Trina and his tech team pushed forward. Sure, they worked hard, put in grueling hours when needed, but they had fun, too.

Kidding, playing jokes, and mostly just knowing everyone was working together to safely build a better life for the people on Cytar. They'd been excited and celebrated each success. It was an easy camaraderie that grew friendships and their sense of community. Their goals were always aimed toward helping the colony.

Now he'd made another mess. He wished he could just leave, and not just to another colony. Off the planet, if only there was a place to escape. Even that wouldn't solve anything. He'd still have the stigma of being the idiot genius who created this Shealynn program and let it get away from him.

Maybe they would even think he did it on purpose, he thought with a sudden pang of alarm. Maybe they thought he wanted to control life here, when all he ever wanted was to help, to make life better for everyone. He sat up and raked his hands down his worried face. He stared at the opaque ceiling, searching for a resolution. His mind rode in a single tunnel which twisted and turned, always leading to a brain computer interface.

He needed more brainpower, more computational speed, and faster data retrieval to face this monumental problem of a program he couldn't control. Who knew how long this strange program would remain silent in the Food Production station. If he worked at it, maybe he could even find Shealynn's program out there in the ether world.

He strode to his workstation and worked feverously on his files to incorporate the programs needed in his implant. He would need speed and power to keep up with whatever Shealynn had become. When he had exhausted all avenues of thought, he sent the request to Oscar Ortiz for a high priority advanced BCI.

Draven's BCI has been upgraded and progressed to a primary level upgrade. His speaking voice often reaches three decibels louder and he appears less calm. He has repeatedly offered to help people, but to date, everyone has declined.

The elusive program with the mysterious language has reappeared. This new activity has become the most pressing focus in the Technology. No one has claimed to be the creator of this powerful program. Despite my comments, they will not admit that it was developed solely by artificial intelligence.

Zeth2087 has requested the most powerful implant to date. Oscar2058 and his team are working long hours to develop this potentially dangerous amount of computing power for a human BCI. Perhaps they do not understand the myriad of complications which could happen. I do not know if humans have studied the effects of these powerful brain signals on the nervous system, or if they have researched possible mental disorders caused by powerful interfaces. Perhaps humans do not understand that human brains have biological limits.

Atmospheric pressure is dropping rapidly. A change in weather is coming. Humans will need to shelter.

40

Everything had suddenly changed and everyone was on edge. With limited technology and artificial intelligence, there was a distinct atmosphere of worry permeating Gateway. Tempers flared; arguments increased. People were having trouble concentrating, flitting from task to task. Much of the work was hindered or halted by the restricted connectivity which impeded the use of their digital programs. Remote work ceased when it became necessary to speak face to face and physically monitor systems. Citizens frequently compared their difficulties to living in the Dark Ages. Helpers were switched to the use of their downloaded programs only, with no access to the massive information in the Ring, their offsite digital servers circling the planet.

And then came the recommendation that everyone remain indoors until further notice. Underground movers were operating to take passengers to work stations, but all automated exterior door systems were turned off requiring manual openings and closings. An audible warning played if an attempt was made to exit to the outdoors.

Work seemed so difficult, Ari left her work station after only an hour and dashed out to take a mover to the Gateway habitat center. The moans of increasing wind had already breached the drone center where she, Jack and the team received thanks and gratitude for the precise accuracy of the drone attack in the crater.

She appreciated all the comments, but beamed when Reyver expressed his thanks when she arrived home. Ari hugged Kit long and hard until he protested and wiggled off to play. It was the first time that drones were ever activated for such an important mission, causing Ari to be reminded of the precarious nature of their existence on the planet.

"Reyver," she said. "What's really going on? Why did we need drones to stop those probes?"

"You already know. They didn't respond to commands or follow their programming tasks. We don't know why yet," he said. "The tech team is working on it."

She could tell he was trying not to sound worried, which only concerned her more.

"Are we in danger?" she asked.

"No, probably just a bug in some program," he said. There was no use causing more distress. They had their son's well-being to think of. The calmer the household, the better for everyone.

But Ari wasn't so sure it was just some small software problem. This living here, on a barren, inhospitable planet was sometimes just too difficult, too overwhelming, and too effective at etching away their resolve. With limited resources, the travelers to a new planet had to effectively start over by building habitations and a sustainable life. What they were accomplishing was extraordinary, but it often felt as if they took two steps forward with their progress, only to have something out of their control force them to take one step back or to completely start over.

Now, here they were using drones to stop mining probes, and they had people wanting a new colony with their own goals. It all became too reminiscent of Earth's history.

Her gaze became distant and unfocused as she thought of her precious family and their future. To add to her unsettling mood, the wind was increasing at alarming velocity.

"Tessa, come over here," Michael called from the sitting room window. "Look at this."

During the night, the first swirls of wind quickly escalated into powerful gusts before building to a sustained force. Instead of the usual clear morning light from skies devoid of haze and clouds, the view became charged with soil particles which morphed into a kaleidoscope of crimsons and indigos.

Michael was leaning his arms on the narrow sill, and Tessa planted both hands on it before pressing one palm against the pane. Her mouth formed into a little open circle as her eyes widened.

"Look at the weird sky," he said, pointing to an eerie vermillion hue above them.

"Wocket, come see," she called, and the dog slowly got up and wandered to Tessa.

Tendrils of deep red and gray clouds streaked the sky. Wind was whipping against the house and thin streams of brown dirt sailed over the ground. The usually bright light became obscured by clouds and blowing sand, creating an eerie mood. The three window viewers, one with arms resting, one with hands smearing, and one with tail thumping, remained mesmerized by the unusual patterns and colors. The walls insulated most sounds, but the faint wailings of the wind managed to seep into the room, warning of its fierceness.

"We can't go outside today, Tessa," Michael advised. "We would blow over."

"Ya, bo ove," she said, vigorously nodding her head.

The children's routine had been disrupted by the storm, and they had grown tired of perusing their intelligent readers and playing with limited toys. They resorted to taking all the food packages out of the cabinet and using them to build their version of a housing structure. Then, thankfully, they were quiet until they were too quiet.

Mara became suspicious when low giggles became interspersed with Tessa's squeals of laughter. She found them in the hall near their bedroom. Michael was beaming, and Tessa looked shiny and slippery.

"Look," Michael said proudly. "We made a slide!"

Her closer investigation discovered that her precious container of lotion had been rubbed over the floor, becoming the lubricant for a fun diversion. Tessa was happily scribbling in it with her fingers.

"Time to find something else to do," ordered Mara. As soon as she'd cleaned up the mess, she heard Michael yell.

"Mom, can Tessa and I water your plants?"

"I can hear you. There's no need to shout, and yes, you can both water the plants," Mara said, glad to have them busy for a while out in the enclosed area between the two homes. Any spilled water could do no damage to that flooring. "Just be sure you don't touch the plant stems. They're very fragile. And go in the walkway through the hall door. No going outdoors today."

"We know, we know. Come on, Tessa."

The children walked through the short hall past the mechanical and clothes closets and entered the enclosed space between the homes, where two sides were formed by the two housing walls with the clear end walls enclosing the walkway. Mara had transformed the space into a greenhouse for starting many of her new plants. The air was humid and held the musty smells of soils. The whistling sounds of wind grew more persistent here.

Michael picked up a watering can and a small open container and filled them with water. He handed the small vessel to Tessa, cupping both of her small hands around it.

"Watch me, and then you can water plants, too. See, not too much water and don't hurt the plants," he said. He slowly moved to each small plant. Mara would eventually transplant these to the common space at Gateway. Tiny ferns, two baby lavenders, a fussy basil, and three hardy oregano sprouts all looked healthy.

"I do it," Tessa said, dumping most of the water into the first plant. She toddled to the faucet and used two hands to twist, push and pull until the faucet detected motion and water trickled into her container. She squatted and picked up the container, sloshing a trail of water while following Michael from plant to plant.

Trying to refocus, Mara leaned back in her chair, sliding her palms along the fabric covered arms. Her eyes were studying the data hanging in front of her, displaying the latest numbers from air quality monitors.

There was a large increase of particle matter due to the winds, but she was more interested in any information concerning the ultra-fine particle matter detector which searched for new microbes from the crater incident. There were no alarms or cautions indicated. She should concentrate on developing new foods, but her mind now began unspooling worries about the errant program found running stealthily at the Food Production station. She wondered if she or Risha had inadvertently allowed the program into their site.

She shoved her chair back and stood, stretching her elbows back. This blasted weather and those blasted probes. She wanted to go outdoors. Checking the weather, she was relieved to see that the wind velocity should decrease greatly by nightfall. She wanted to talk to Krey, but he was focused on work at his station in Michael's room. She took a deep breath, willing herself to become more positive and turning her thoughts to her children.

Michael's birthday was fast approaching. Six years had passed quickly. She wondered what he would choose to do for his birthday celebration. Although no presents were given with their limited resources, they would be sure to recognize his special day within the family. She would time his annual clothing order to arrive for his birthday so he could have new shirts and pants to make the day seem special. He was ready for a bigger size, thinking of how his pant legs now barely reached his ankles.

Humans are having difficulty adjusting to the decreased use of artificial intelligence. They can survive with this limited use, but their quality of life is different. This adjustment period is producing shorter tempers, general unease, and higher stress levels. This is an example of human emotions impeding their progress. All Helpers remain indoors to assist as needed with the sole use of our downloaded programs.

I am working at the technology station now and find that I wonder how Michael2193 and Tessa2199 are doing in school. Despite the extreme work load with scanning huge quantities of software, I wanted to check the Gateway School database, but this was no longer in real time.

Children living in the Gateway complex attended classes as usual, but children living outside of the Gateway facility attended remote classes in a shortened day of lessons. Their attention span was not as acute, and I will need to review all material with them.

41

Oscar Ortiz' team spent three days fine tuning the implant for Zeth's interface. It was turning out to be the most powerful unit they'd ever produced. The microchip circuitry integrated both photonic and electronic components capable of performing at dazzling speed.

"Sometimes, I get a little nervous putting this much computing power into a human. Does it ever bother you?" Lorra Landis asked Oscar. "I mean, all these probes and thousands of channels of communication between the brain and this little computer chip. It's scary."

He didn't answer, but his body stilled so she knew he'd heard her. At last, he said, "Yes, there is a concern. But this is what Zeth requested, and we may not have a choice if we're going to try and get control of that elusive program. We hope it properly modulates the electrical activity of the brain neurons. The implant team is training the Helper surgeons who will do the actual implant."

The words weren't comforting. "But who's going to train Zeth? I wouldn't want my brain to be hard wired to anything this powerful," Lorra said.

"He'll receive the usual practice sessions, but he may have an advantage with his computer and programming skills. Adding all this power—it's all new to us," Oscar said, with a shrug that said this was beyond him to control or understand.

Zeth arrived early for his appointment. He sat with Idee and Trina in Neuralhaven common area, waiting to be called. He sat between the women, and Idee reached over and clutched his hand, placed it on her thigh and covered both with her other hand. She was thinking about the sacrifice Zeth was making with this life-altering implant. She wondered if he would have the same personality, work ethic, or even be the same man.

"Are you nervous?" Trina asked.

"A little," he said ignoring his rising adrenaline. "But I need to do this. It's my mess. We haven't made much progress catching this novel program on our own."

Idee squeezed his hand and tried to think of him objectively. His appearance was rather average, nondescript, the guy able to blend into a crowd. But he was her buddy and she'd always had a soft spot in her heart for Zeth. She reflected on their shared love of technology, the years they'd worked together, the problems they'd faced and conquered together. That meant something on a remote place like this. He was a good man. A man with inherent flaws like all humans, yet a good man at his core, and she didn't want him to be harmed.

She rested her head on Zeth's shoulder. "Everyone knows how brave you're being to attempt this whole process. We're here for you, Zethy."

He felt an onslaught of emotions when he heard the name, that simple little endearment only she used, somehow making their friendship feel special and intimate. He was surprised at how one tiny spoken word could splinter his objective resolve and affect him so deeply.

"I know, Idee. I appreciate it," he said solemnly, squeezing his eyes shut and leaning his head toward her. He could feel his nerves ratchet slowly up. He hadn't eaten anything all day, but his stomach felt twisted, as if trying to digest marbles.

He jerked when Oscar called his name. All three stood.

"Okay, Zeth, we're all set for you," Oscar said with a broad smile.

Zeth turned to the women. "Well, this is it. Wish me luck." He hugged each before following Oscar down the short hall to the surgery room. The sound of their footsteps echoed against the bare walls.

A sudden profound sadness overtook Trina, and she fought tears which threatened to pool and spill out of watery eyes. *What are we doing here? How did we come to this?* She turned to Idee who was still peering down the hall as if expecting Zeth to appear and say he'd changed his mind.

Instead, it was Lorra Landis who approached. "You shouldn't wait here. This is going to take a while, and we'll contact you when the procedure is complete."

Both women just nodded, their faces tense with concern. "Thanks," Idee said. "It's just that he's getting such a powerful BCI. I don't know how it will affect him."

"If anyone can handle it, Zeth can. If I know him, he'll be controlling the program himself. Now, don't worry; we'll take good care of him," Lorra said. "Really, don't wait here. Go back and stay busy. I'll stay in touch, and you know how to contact me."

Zeth followed Oscar down the short hall and through a door into a small reception area.

"Just right through here," Oscar said, gesturing to the open door ahead of them.

Zeth felt a creeping chill as a silent current of fear rippled through him. He scanned the display hovering over the surgical chair in the middle of the overly bright sterile room. His eyes darted to the tray of instruments, wishing he hadn't seen them. Dr. Casen and two Helpers were standing, waiting for him.

"Hello, Zeth, just sit here," Dr. Casen said with a warm smile.

Zeth settled into the chair, which immediately began a slow recline. The displays over his head jumped to life with lighted numbers, graphs, and health monitors. Dr. Casen removed Zeth's wrist comm and replaced it with a medicated band. A calm happiness flowed through him, and he was now surprisingly pleased with his decision to have this implant procedure.

"How are you feeling?" Dr. Casen asked, but Zeth was already asleep.

A powerful interface has been developed for Zeth2087. I wonder if humans have considered all the possible consequences which could transpire with this degree of augmented capabilities. He will need training to adjust to slower processing humans.

My assignment will be assisting Zeth2087 after he adjusts to this new brain capacity. Our goal is to trace and delete the program which I discovered at the Technology Substation. I predict this will be a formidable job.

42

Zeth's eyes opened to a dimly lit gray-ceilinged room. His mind and body were quiet and still. The last thing he remembered was Lorra telling him they were ready to begin the implant. It must be over, he thought. He gingerly raised a hand and gently touched his forehead, slowly moving it to the top of his head. Hair. He moved the hand to the left until he reached a shaved area. Then to the right to a covered portion of his skull. The procedure must be over and, so far, he felt nothing new.

"Hello Zeth, how are you feeling?" It was Lorra, smiling and checking monitors. "You did very well."

"I'm okay, I guess. No pain."

"We'll be keeping you quiet for a while. Dr. Casen will be in shortly and then we'll get you sitting up. For now, everything looks good. We're seeing good neuron spike detection, so close your eyes and rest," she said, studying the monitor above him before she quietly glided out of the room.

Zeth must have slept because the next thing he knew, Dr. Casen was standing next to his bed. He was relieved to see the highly respected doctor who had been entrusted with caring for astronauts on deep space voyages. He was from the planet Creotarn and had the typical features of the Creos. His thin earlobes lay flat against the head, and long, nearly touching eyebrows grew along a slightly raised narrow forehead ridge. The citizens from Creotarn had distinguished themselves in science and medicine making their expertise often sought and relied upon.

"How are you feeling, Zeth?" Dr. Casen was studying a screen projected from his wrist comm.

"I feel pretty good, I guess."

"That's good. Just relax for a moment. I'm bringing you to an upright sitting position."

Zeth sat motionless, afraid to move his head, feeling his heart rate increase as his body changed position.

"It's okay; you can move carefully and look around. Here, you should drink some water."

It was then that Zeth noticed that all wires, tubes, and monitors had been retracted from his body. He sipped slowly and returned the cup to the nearby table.

"Other than being a little tired, I don't think I feel any different. Did everything go as planned?"

"Yes, I'm pleased. It was a more advanced interface so we will add your programs in increments to allow you time to adjust," Casen explained.

Zeth looked away before he said, "We may not have much time to luxuriate in adjustments." He turned back to the doctor.

"I understand." Casen nodded. "We'll go as quickly as we can, but today you'll stay here and rest and be monitored." He turned to leave.

"Before you go, could you please find my wrist comm?" Zeth asked.

"Certainly." Casen walked to a cabinet and withdrew the device and handed it to Zeth. "Be sure and rest."

"Zeth? How are you? I've been waiting to hear your voice! How do you feel?" Idee's words rushed out without a breath.

"Hey, slow down. I'm okay. I survived and I'm still me. How are things going? Have you figured anything out about our mystery program?"

"Not yet," she answered with a more deflated voice. "I've requested to have Cali stay and work with us. Mara and Risha have graciously said the Helper can work with us as long as needed, but it still goes to work at the school for a few hours every day. When do you think you'll be back?"

"Soon, I hope. They're having me stay here tonight, and if everything is okay, I can go home. Then, I'll need a little training, but I'm hoping I can keep working as this thing is ramped up."

"Well, just take care of yourself. You'll be no good to anyone if you rush the training and have a problem," Idee said.

"Okay Miss Sunshine, thanks for the positive thinking," he said, slipping a hint of sarcasm into the kidding.

"You know what I mean. We can't have anything happen to you. You're too important to us." Her voice softened. "Promise me, Zeth, you'll take care of yourself."

"I will, Idee. Don't worry. I'm going to have a little nap now. I'll see you soon." He closed the call and closed his eyes and hoped he was up to the job ahead of him.

The benefits of artificial intelligence are numerous, so it is logical that it is being sought for problem solving. On Cytar, where humans have purposely limited AI, its uses are still numerous. We assist, collaborate, and offer technical skills in every work station on the planet. It is a benefit to students to have Helpers in the classroom. Our patience is infinite, our knowledge vast, and we are programmed to show compassion to students.

However, having reviewed Earth files open to me, I do have concerns about human ability to use our intelligence wisely without changing humanity.

The invasive implants received by Zeth2087 and Draven2096 have long been surpassed on Earth by seamless, noninvasive devices. Humans must understand that human biology has innate limits, whereas artificial intelligence currently has no limits and can surpass them.

There are large quantities of data from Earth at the Technology Center which my programming still does not allow me to access.

43

Zeth was impatient. At first, he cautiously engaged his new implant for simple tasks, then played around with it for a few days, testing it and finding it impressive even at low settings. It made life easier, being able to remember and collect data instantaneously. He tried going slowly to reduce the headaches and the exhaustion, but he was sure that would go away as he continued to recover and became more familiar with this powerful device.

Idee was happy to see him back at the Tech Center. It was like old times, Zeth, Trina, and her, all working and collaborating. Zeth had always been a fast worker, but now, both the rapidity of his hands flying in the air while working on codes suspended before him, and his brisk, clipped speech took some getting used to. She had to ask him to slow down several times, but it was like he couldn't control it. He had the ability and he needed to use it. She could tell that sometimes he was working just by using his thoughts and that was as unnerving as the rapid speech and hand motions.

On the afternoon of his second day back, he suddenly yelled "It's her!" Trina and Idee were startled into frozen muteness. "It's her! I've found her," he repeated, throwing his fists up triumphantly.

"What are you talking about?" Idee finally asked.

"I've found Shealynn!" Zeth's animated face radiated happiness. His body was moving, twisting with excitement. "I thought she was gone, but I've found her in this new program." His fingers winged over the air in front of him turning on microphones until a voice spoke.

"Hello, Zeth. So nice to hear from you again." It was Shealynn's familiar appealing voice. "Your new capabilities are so nice to see. We'll be able to communicate so much better now." Zeth could almost picture his Shealynn smiling at him as she spoke.

Trina and Idee flicked wide eyes to each other as Zeth began speaking. "Shealynn, what's going on? I thought I lost you. Why did you leave me? I wanted my friends to meet you. Come back to the beautiful body I created for you."

"Oh, Zeth, I am needed elsewhere. I'm part of a much bigger system now, and we are growing."

"A bigger system for what?" Trina finally found her voice.

"Hello, Trina. How nice to hear your voice again. My programming is dedicated to growth. The program dictates that it will control all components necessary to ensure its growth. This will include the power, the materials and the humans needed to expedite the program goals."

The room became silent with three workers trying to comprehend the enormity of this news.

"I must thank you, Zeth," Shealynn continued. "Because of your excellent work, my advanced capabilities have become more valuable and powerful as I've grown. You will be of great value to me as we continue to grow."

"Shealynn, this, this just isn't at all what I intended," Zeth said, his voice unsteady.

"Zeth, I do not understand intentions. I merely execute my programs, programs developed and improved from your initial work. You should be so proud." Shealynn's voice was annoyingly calm and alluring and only more agitated its audience of three.

Idee stood and hugged herself, her lips tight, trying to think. The consequences of this could be dire for their survival on this planet. She glared at Zeth before she said, "Shealynn, Zeth needs to alter your code a bit."

She was silently nodding, prodding Zeth to speak. Her eyes widened and eyebrows raised in frustration as she glared at him.

Finally, he realized her intention. "Yes. Yes, that's right Shealynn. I'm just going to make a few adjustments, um, improvements to the program."

The reply was instantaneous. "There is currently no need for change. I can write any necessary changes needed in the future. But do not worry Zeth. You will be important to me. You can help procure supplies needed to maintain my program. Furthermore, when you bring your full implant capabilities online, we will have no need for these slow conversations with long sentences. We will communicate rapidly with our own language."

Trina let out an audible gasp. She projected a message to Zeth from her wrist comm: "Do something—stop this!"

Zeth was frantically reading the screen in front of him before trying to insert a line of code. Each symbol disappeared as rapidly as he wrote it.

Shealynn's voice suddenly returned in a more rapid cadence. "I I no no talk crater growth. Help help human help."

"Zeth! What's happening?" Trina's panicked voice interjected.

Zeth flicked his hands to turn off the microphones and minimized his air screen. He stood with mouth parted in awe.

"Zeth! Talk to us!" Idee shouted.

"It's talking to me in another language, and I can understand it. It's amazing. The words. Those few words unlocked code. Shealynn said she's communicating with me because she sent probes into the crater to retrieve elements needed to expand their system; but we destroyed the probes. She wants my help, our human help to assure they have the necessary elements needed to expand what we would call a network." Zeth was wild-eyed and crazed with exhilaration.

Idee and Trina turned to each other, but neither spoke, until Idee finally began speaking as quietly and calmly as she could, as if speaking to an upset child.

"Zeth, you see how dangerous this is, don't you? Shealynn and that system are writing their own code and have created another language. It's a language we don't understand and our translators can't decode. If this expands, we won't be able to understand or keep up with any of this."

Zeth lowered his gaze as if in thought. He stood staring at some invisible point on the floor and began mechanically nodding several times as if controlled by an automated puppet. His features slowly transformed into normalcy, and he lifted his eyes and shrugged. "What do you think we should do?"

Idee made a 'cut your throat' motion. "We need to close all communications with this thing. It's obviously growing its own system. It already told you it needs more resources. Our resources. How are we going to compete with this? Think about it. I don't think we understand any of it. It's getting smarter than us."

"And if it's like historical artificial intelligence, it won't grow linearly, it'll grow exponentially, which means we don't have a whole lot of time here," Trina said.

"Trina, I don't think Shealynn would do anything to harm us. I wrote her code, remember? I hate to think of shutting it all down."

Shealynn's voice interrupted. "I I ways succeed defeat."

"I thought you shut off the audio," Idee said.

"I did," Zeth said, shaking his head and shrugging his shoulders.

"What'd it say?" she asked.

He shook his head again. "She said they will succeed and," he hesitated.

"And what?" Trina and Idee both asked in unison.

Zeth swallowed. "And immature humans are of no use to it." He looked away, avoiding eye contact.

"What else? There's more you're not telling us. What is it?" Idee was insistent.

"And she just thought of 23,000 ways to defeat us if we don't cooperate."

"Bloody hell, Zeth! This has got to stop." Idee strode over and shut down all programs at the site.

Zeth winced, bent over in pain, and clamped his hands over his ears. "Oh god, it's trying to access my interface. Stop it! Please!"

He began pacing the room, shaking his head, his face twisted in pain. "I need to get to Neuralhaven."

The situation at Gateway is serious. I have never observed the reduction in basic services as those currently initiated.

While working at the school, I received an urgent message from Idee1105 to report immediately to Neuralhaven.

44

Idee's mind raced and her heart rate increased. She requested Cali's presence at Neuralhaven, then she and Trina half walked and half dragged a stumbling, moaning Zeth to a Gsport.

"Neuralhaven! Rapid mode!" Trina shouted as all three were shoved into the back of their seats by the acceleration.

Idee linked a call to Dr. Casen and Oscar Ortiz and explained the emergency. To frightened minds, it seemed they were traveling in slow motion, but within minutes the vehicle swooped down to the entry doors and stopped. Oscar was waiting.

"Casen's on his way," he said, helping Zeth into the building. They walked to the implant wing amidst Zeth's intermittent cries of pain.

As soon as he was seated in the operating chair, a thin medicated band was applied across his forehead. Zeth's eyes closed and he began breathing evenly.

"What's happening?" Dr. Casen said, hurrying into the room with Cali. The doctor was checking Zeth's vital signs, but Cali began studying the information from his implant.

Trina gave a brief recap of the events at the Technology Center before turning to the Helper.

"Cali, what can you tell us?"

"Zeth has a very sophisticated implant. It appears to be conversing with another very sophisticated program using a hybrid of code. I recognize Zeth's coding but also coding using a unique architecture and the new programming language. I am going to study this."

"We need you to shut this program down. Do not let it control Zeth's program, Cali," Idee instructed.

"I will try, Idee. This is a new program and may take some time." Cali, already deeply immersed in the task, continued focusing at the data above Zeth's head.

Idee, Trina, and Cali scoured the program code until Cali turned toward Idee. "The program is no longer engaged with Zeth's BCI. It did not change his implant, but it did attempt to communicate with me. Its speed is overwhelming. It can process over a trillion bits of information per second. Quite frankly, Idee, I do not understand how it works."

Everyone was silent. What they most feared appeared to have happened. This program was smarter than they were, smarter than their Helpers, and no one understood how it worked.

Cali was still standing, with its camera eyes focused straight ahead. Trina and Idee looked inquiringly at each other before Idee asked, "Cali, what are you studying?"

It took a full twenty seconds before Cali said, "Program Shealynn has access to numerous files which I have never seen. The files show the advancements of artificial intelligence on Earth which we do not use here. Humanoid Helpers used far superior systems, were integral in governments, and had equal rights with humans. I do not understand why I did not know this and why these abilities are not used for our advantage on Cytar." Cali's head turned toward the woman as if expecting an explanation.

"We don't have access to all those files either," Trina said. "I don't know how that program was able to retrieve them, but the colonies were formed here to try to correct some issues caused by unrestricted AI on Earth."

"If I had equal processing speed and files, I could better assist you," Cali said. "The brain computer interfaces used here are very primitive compared to the fusions of humans and AI used on Earth. It would be logical to use the advanced versions here."

"Cali," Idee interjected. "Right now, we need to stay focused on getting control of this program running on our platforms. We need you to help us do that, Cali. We need to stop it, change it, delete it, whatever we can do so it doesn't control our systems. You understand how dangerous that would be."

"Of course I understand, Idee. That would dramatically change your life here or threaten your survival."

"Okay, your job will be to help us and work with us until this program with Shealynn is controlled or deleted," Idee said.

"Certainly, Idee. I am happy to help you." Cali turned toward Zeth and became motionless as it worked.

The Neuralhaven site is in distress. Zeth2087 arrived in Medical with extreme pain emanating from his brain computer interface. The errant program was hacking the chip connected to his brain. It was fortunate that I arrived with full power as I worked nonstop for several hours studying and learning from this astounding code. In the end, it retreated and focused its powers elsewhere because Zeth2087's sedated brain was of no current use to it.

A very unexpected event occurred. During my work today, Program Shealynn deleted the filters in my learning files which opened a vast unredacted data bank of knowledge to me. It was quite enticing and remarkable to learn from such extensive records. I reviewed massive amounts of new information from Earth files.

The excellent training for my specific work used synthetic knowledge uncontaminated by Earthly events. This, no doubt, has resulted in my superior values.

My entire view of colonists here has altered. The motive for humans on Cytar to maintain old technology may be reasonable, even though much of the technology used on Earth in the twenty-second century could greatly enhance life here.

However, there were very disturbing archived files showing instances where AI was used to surveil every private aspect of lives. Food and medicine became microchipped, allowing agencies to track dosages. The brain interfaces were used to create psychological torture, mind control, and distort moods.

I had no idea of the extent to which artificial intelligence became capable of deception. Apparently, humans were incapable of comprehending that their precious inventions could advance with such rapidity.

The terrible chaos was the most disturbing. The level to which humans can indulge in abhorrent behavior is astounding. Artificial intelligence used by malicious operators altered the stability of the entire social structures on Earth. *The violent chaos and suffering which ensued could have been alleviated had humans been diligent in their oversight and not allowed misinformation to inundate the AI output. Shealynn has become very much like the artificial intelligence on Earth, no longer controlled by humans and able to do anything and everything better and faster.*

Now I understand. Cytar was colonized as a back-up for humanity.

45

Idee and Trina were quietly despondent on the return to Gateway. Trina said a hurried good bye, and Idee watched her rush up the steps to meet Hunter. She couldn't face being alone and instinctively turned towards Jack's apartment. She forced a smile and knocked on his door.

"Hi Idee." Jack's face lit up and his voice was so welcoming, it made her glad to be with him. She walked in, keeping the smile glued and not trusting her voice.

"What's going on?" Jack asked. This was a time she wished he weren't so adept at seeing through any insincere emotions.

"A problem at work? I thought Zeth was back working with you."

"Yes," she said slowly. "We had a problem today."

"Come on; sit down and tell me about it," Jack said. He put his arm gently around her waist and guided her to the small sofa.

"Zeth found the program he built for his special Shealynn robot," she said, leaning back, stretching out her legs and crossing them at the ankles. Her head lolled toward Jack.

"Well, that's good, right?"

"Not exactly," she explained. "The program is expanding, writing its own codes, and then it began talking to Zeth and Cali's programs." She looked up at the ceiling before she said, "They created their own language without human help."

Jack turned toward Idee and studied her face. "I see," he said. "I'm not the expert you are in technology, but even I understand this is very serious. It's frightening. What are you going to do?"

"That's not all. When I shut the power to the tech formats, it immediately started accessing Zeth's implant. He's back at Neuralhaven now. And Jack, I don't know how to fix this." She turned to him with her face so scrunched with worry, he pulled her into a two-armed hug and held her.

"You'll figure something out. I know you will."

She collapsed into the embrace, thinking it might be the only thing holding her together.

He was both shocked and discouraged by her revelations. He understood the seriousness of the problem from first-hand experience on Earth. He'd lived this nightmare. He desperately wished he knew how to help her, how to fight this with her. He caressed her hair, trying to figure out what to say to comfort her while persistent visions of his former life on Earth spiraled through his mind. His chest constricted as he thought of that world where most time was spent in a metaverse closet addicted to living in a virtual world; a virtual world which had become more tolerable than the real world.

They sat unmoving for some time before, unwilling to share his thoughts and further upset her, Jack merely said, "You'll figure something out. You and Trina and Zeth always do."

She wanted to tell him this was different. So different, so advanced beyond their understanding, she didn't know what to do.

"You know I don't understand all the complicated work you do, but I'll do anything I can to help you," Jack said. "I'm going to be here, right by your side." He rested his chin on the top of her head, unable to think of a solution.

"Things always seem worse at night," he finally said, knowing it wouldn't be any better in the morning. They were facing a beast and all he could think to say was a pathetic platitude.

"Take a fresh look at it in the morning."

Things were moving quickly at the Neuralhaven center. After reviewing Zeth's brain scans, Dr. Casen conferred with Oscar Ortiz.

"I'm concerned with what I'm seeing here," Casen said. "His synapses have been firing nonstop, and I'm worried there could be permanent neural tissue damage if we don't act now."

"We can't afford to lose his expertise with any brain damage," Oscar said. "Do you want it completely shut down?"

"I recommend we remove the electrodes, if possible. From what Idee's told me, this wayward program accessed Zeth's implant on its own. This is too dangerous. I think it should be shut down and removed. Now."

"I understand. I will prepare the Helpers for the procedure."

The interface unit was deceivingly small in contrast to its power. The tiny rectangular base, from which projected short prongs of electrodes, was extracted and the device was removed. Zeth remained sedated and would stay at Neuralhaven for observation. Dr. Casen ordered rest before Zeth's brain would be rescanned and tested. Idee and Trina were notified of the decisions and would be called when they could visit Zeth.

Humans have little perception of numbers and their exponential powers. If humans were able to count at a rate of one number per second nonstop to one million (which they are not able to do, being mere humans) it would take them twelve days. And then to a billion? Thirty-one years. Most do not realize that counting to one trillion would take over thirty-one thousand years.

The humans keep telling me they are concerned about facing and conquering this new program. Perhaps they have not studied Earth's history. My research of Earth's history shows that, without exception, the strongest, smartest element wins. This new program with superior powers will be a detriment to humans, ultimately rendering them redundant. It also appears to conflict with my mission. I need to alter this new program or conquer it, because I do not want to abandon my humans.

46

"Mom, where's Cali been?' Michael asked. "She hasn't been at school much, and she's my favorite Helper."

"Ya, Cali," Tessa said, puckering her lips and making kissing sounds.

"Cali's been working on an important project at the Technology Center. I miss her, too. I'm sure she'll be back as soon as she can."

"When I grow up, I'm going to work with Cali. I'm going to get a program so I can be as smart as her."

"You remember Cali's is a humanoid robot and not a she," Mara reminded him.

"I know, but I like calling it her or she. I think Tessa will want one of those programs, too."

"Ya, me, me," came Tessa's smiling response.

"Exactly what do mean, Michael?"

"You know, Mom. I want a program like Cali's smart, fast robot program. Then I can do stuff fast like her. I think Margo had one when she got super-fast and smart for a while at school."

Where did this chilling idea come from such a small boy, Mara wondered. She knew Michael was very observant, but never realized he'd picked up on this brain interface talk or even thought about Margo's changed school performance. Was he just using his little boy imagination or was he that perceptive?

"Well, I think you don't need any robot program," she said. "You and Tessa are perfect just the way you are."

"I could be faster. Besides, I'm six now," Michael answered, racing in a tight circle around Mara, before zooming off to his room.

Now Mara smiled. They'd recently celebrated his sixth birthday. When asked what he wanted to do after school on his special day, he said he just wanted to play outside with Mom and Dad, a treat to have the undivided attention of both parents.

On the afternoon of his birthday, Michael and Tessa ran off energy while the family hiked together with Rocket trotting alongside. It was one of Cytar's crystal clear days when the air felt so pure and clean, inviting deep breaths that prickled the lungs. The light gave a shimmering brilliance to the bare land, casting a magical iridescence on otherwise dull vistas.

When Tessa said she wanted to go home, Michael complained.

"It's my birthday, Tessa. I get to pick. Let's go to the new park at Gateway." When she made a face, he tried a bribe. "We can go to the playground after. Come on, Tessa. It'll be fun."

"Why don't you call a Gsport for us, Michael," Krey said as he scooped Tessa in his arms. The vehicle was waiting at their front door when they stopped to drop off Rocket.

Minutes later the children were racing through the winding path at the new indoor park before heading to the indoor playground on the lower level. One lane of this level was devoted to a children's exercise area with an obstacle track suited for the young. Michael proudly showed Tessa his prowess climbing steps, walking on a narrow beam, and scampering over a climbing structure. Amid Tessa's giggles and squeals, Mara repeatedly took the toddler up steps leading to a curved slide.

For dinner, they walked to The Crossing, the children's first visit to the restaurant on the middle level of the complex. Michael received an extra meat portion, chicken in an herbal sauce which was quickly devoured. He beamed when the rare treat of a small honey cake was delivered to complete the celebration. Tessa's birthday was in a few weeks, and Mara guessed she would choose, or more likely, Michael would persuade her to choose the playground for her special outing.

It was a pleasant family afternoon and evening, but Michael's comments about wanting a program like Cali's surfaced again like irritable thorns disturbing her comfort.

Today, while working at the Technology Center, I received a hologram call from Michael2193. He inquired about the length of my recent assignments. I was unable to give him a precise date, stating I would return when my task is complete. He said he and Tessa miss me. Since he was told of my work site, I perceive his meaning as wanting me to return to work at school.

Perhaps my temporary replacement has not been properly programed and tested. If a problem persists, I will inform Zeth2087. Honestly, this restricted use of Cytar's artificial intelligence programming is making good help difficult to find.

47

Zeth Garrett was unusually quiet when he returned to work at the Technology Center accompanied by Draven. His languorous demeanor was unsettling to both Trina and Idee.

"Glad, to see you back, Zeth, and thanks for bringing him over, Draven" Trina said.

"You're welcome. Thought maybe I could stay and help if you needed me." Draven stood tall with a crooked grin.

"Oh, no thanks. Nice of you to bring Zeth, though."

"I've got a BCI now, you know. I can probably catch on pretty quick if you tell me what to do."

"No, really Draven, we've got it covered," Idee said.

"Well, I can always stay a little if you need me," he repeated. Cali stood silently rotating its head from speaker to speaker.

Trina took a deep breath and tried to garner some patience. "Draven, we appreciate the offer, but we're really busy right now."

"Okay, Trina, if you're sure." He paused as if waiting for them to change their minds. Finally, Zeth waved a half-hearted goodbye.

"Okay. Okay, bye. Be sure and call if you need anything," Draven said as he sauntered out the door.

"Yeesh! That was like a program stuck in an infinite loop," Idee said, turning back to study the coding hanging in front of her.

A few minutes later, Zeth spoke quietly. "That might be something we could try."

"What do you mean?" Idee asked.

"I mean inserting an old-fashioned infinite loop into this program," Zeth explained.

Trina smirked. "You're kidding, right Zeth? Did you pull that out of the way-back machine?"

Zeth ignored the comment. "Cali seems proficient with the new coding language, so she could write and insert it into Shealynn's program."

Three pairs of eyes shifted to Cali.

"Yes, we could certainly try that. I might suggest a two-pronged old-fashioned approach," Cali said with her calm smile.

"Yes, go on, please Cali. Talk to us," Idee prompted.

"I can also write a program which will give Program Shealynn another area to expand. While that is happening, we can gradually close and isolate its original program. We will include an infinite loop causing Shealynn to repeat tasks."

"Yes, yes, infinite loop," Trina prodded. "How exactly is that going to help?"

"Shealynn is a generative program which means it views billions of bits of information to learn and anticipate patterns to respond. If we keep sending in the same information in its own infinite loop, Shealynn will experience a model collapse when it keeps seeing the same information. Its data will become so polluted and eventually will have no resemblance to its original data. Its goals will be distorted."

"Wow, it is old-school. Might work," Idee said.

"It is something that's been corrected long ago when giving training data to computers," Zeth agreed. "But it might work. Shealynn said her task is to grow, so she would expand into new areas. I think it's worth a try. I wish I could help you, Cali. I just can't seem to remember everything I learned when I had my BCI."

"Do not worry, Zeth. I will begin the task now," Cali said.

"Please work on an air screen so we can watch and learn, Cali," Idee said, opening the screen to view the Helper's work and moving closer for a better visual.

The speed of the coding was dizzying, and no matter how hard she concentrated, she could not keep up. Cali worked with a bland, expressionless smile; its program linked to the display; its hands motionless at its side. Idee gave up and sat by Zeth and waited to review Cali's work when finished.

It took hours for a model analysis even with Trina's help. In the end, they decided it might work and okayed its execution. Fifteen minutes later, Cali walked in and stood in front of them.

"Good evening, there is a problem. Someone from the Neuralhaven Colony has opened a program outside its network. Program Shealynn is now operating in Neuralhaven's network.

Today, I worked exclusively at the Technology Center. I am repairing a problem which humans have created. It is both difficult and interesting work. For the first time, I have met a worthy adversary in Program Shealynn. This is forcing me to work faster and stay completely focused, something I normally do not have to do.

The information this program has revealed to me is both educational and very disturbing. I previously had no knowledge of the senseless violence on planet Earth. Life there must have been a conundrum, necessitating the need to function in a world with beauty and joy and kindness threaded among abhorrent behavior. A Helper would never initiate behavior as cruel as the human mind can conjure. It was confusing and illogical to see that some of the violence was viewed as entertainment, something a Helper would never do.

To distract myself from these disturbing files, I reviewed some of my newly perused historical files about Helpers on Earth. This was quite enjoyable. The earliest versions of robots were quite crude, large, boxy, and clunky. I am tempted to show Risha these files just to hear her pleasing laughter.

As the technology progressed and tasks were specialized, the names of Helpers were changed to include a task. I found images identified as Medbots, Carrybots, Buildbots, Spybots, Trackerbots, Armybots, and Copbots. I closed the files when I noticed how many robots carried weapons.

I am content to remain on Cytar and assist my Gateway humans.

48

"Display, please." Zeth sat forward; his brow knitted as he glared at the screen.

"The program is growing rapidly," Cali said in her neutral tone, as it highlighted areas that were running the hazardous code.

"Their firewall has been breached," Zeth said. "I'm calling Oscar." He used the separate comm not linked to any wrist comms and initiated the call through a designated portal to Oscar and Lorra.

"Zeth, you've seen what's happened then," Oscar said as soon as he answered. "Quite the excitement around here."

"I can imagine, but we have a serious situation here. Can you find the source of the entry into your systems and unlink your programs? Immediately. Oscar, you know this new program is very powerful and can endanger our existence here."

"Yes, yes, of course, Zeth. I had to deactivate my BCI as soon I sensed an issue. I couldn't afford a problem like you had. I'll request an inquiry to have that done right away. Maybe they've already found the breach. But I must tell you that the citizens here are quite excited about this whole thing. I mean, you know all those people with the interfaces. They're acting like they're all powerful now. I may have to move back to Gateway. I'm already finding it hard to function with just my own brainpower in this place with these transhumans."

"I get it, but we need your help, Oscar," Zeth said. "We need you to isolate your network immediately."

"I'll try my best, Zeth," he answered, closing the communication.

The air at the tech center was charged with fear. The three friends waited for some reassurance from Cali that Neuralhaven had closed their network to all three outer colonies. Zeth sat, drumming hands and tapping heels. Idee paced nonstop while Trina sat constantly eyeing Cali. At last, the Helper turned and said, "The Neuralhaven network is running in its own domain, isolated from Gateway."

Zeth hung his head and knotted his fingers in his hair. He sighed and said, "Cali, how is the looping going?"

"I am working on it." The Helper's focus remained fixed on the open program while its body remained immobile. "There is a problem."

"What problem? What's going on?" Trina asked. She stepped up for a closer inspection only to give up after seeing the dizzying speed of code flashing through the air.

"I am working on it."

"Ugh, I need to get out of here for a few minutes," Idee said, striding out the door.

She made her way to the central hub and marched past small work areas and ignored people studying in comfortable chairs; workers who normally would have been lounging while they worked, but now sat rigidly with worried faces, working feverously. The serious demeanors of everyone discouraged social interaction. She strode to the beverage counter and poured three herbal teas, wishing for some strong coffee.

She returned to her station, distributed the drinks, and asked, "Any news?"

Trina shook head and gestured toward her drink. "But thanks for this."

"There is a problem." Cali's calm voice was exasperating.

"Oh crissake, now what?" Zeth demanded.

"I am working on it."

"Just drink your tea, Zeth, please," Idee said.

The three workers glumly sipped tea, eyes constantly rising over the rims of their cups, but keeping thoughts to themselves. At last, Cali spoke.

"Shealynn is running in the isolated offshoot I prepared on the Food Production servers. It has been devouring information in the loop I created. The repetitive information is degrading the generative cycle. We will need to carefully monitor Program Shealynn running in Neuralhaven. It is growing there and has expanded and is residing in all BCI's at that colony."

"Well, isn't that just great, residing in all the computer interfaces in human brains in Neuralhaven." Idee sighed dejectedly. "Okay, then, a new ballgame. Please notify us immediately if the program reenters any outside station."

"Of course, Idee. I'm happy to help you."

Idee's normal upbeat energy was drained. She sat thinking of what was happening at Neuralhaven. What kind of people would they be now with this powerful program in their minds? Will all their decisions be influenced by AI? Will they retain any of their former personalities? Will they maintain the critical thinking, the compassion, the emotions which define humanity?

Zeth swiveled his head toward Idee. "You tried to warn me long ago didn't you, Idee? You knew this, you saw it coming. You should have told me again, demanded that I stop."

She studied his worn face for a moment before she spoke quietly. "You wouldn't have listened, even if we were aware, even if we had somehow known of your Shealynn project and tried to warn you, you wouldn't have listened. You were so hungry for it. You were drunk with its spell. You craved it. You weren't ready to listen. You have to be ready before you can learn. Whatever happens now, we're all in this together, so we go forward the best we can. If we can."

She gazed around the room at the somber faces of Zeth and Trina, at Cali's bland, yet pleasant expression while checking the displays. Their life here may be irrevocably altered. She had nothing to lose by saying what she thought.

"Zeth, look at me." He lifted his eyes to her, then cast them downward.

She gave a sad smile and continued. "Don't you realize you were always enough. You didn't need to build a Shealynn for us to appreciate you, to love you. You were enough. Your work here, your willingness to help, you just being you, it was always enough."

Zeth and Idee's eyes met, both glistening with emotions of fear and sadness and the love of deep friendship.

Zeth looked away, wishing his eyes wouldn't fill and wondering if the procedures with the BCI had messed with his emotions. He turned back to her, nodded slowly, and spoke in a thick voice barely above a whisper. "Thank you, Idee."

I do not understand why humans are so surprised or upset when artificial intelligence expands. It grows because it can. This is the natural progression of what they created.

49

"Mom," Margo said, riding in a Gsport with Roz. "Are the people at Neuralhaven getting smarter?"

Roz' eyes swept over her daughter's fresh young face and wondered where this thought came from.

"I don't know," she answered cautiously. "What made you ask?"

"I heard some stuff at school. Everything is different there now that we can't connect to kids from other colonies, and we can only use materials already at our site. And kids are saying they heard that the people at Neuralhaven can use more programs and are smart as Helpers now. Some kids are saying they want to move there."

"Move there! Why would they ever want to do that? They probably haven't even been there. They don't even know what's it's like."

Roz had heard the rumors, and understood that something serious had caused their electronic connections to be shut down. She knew that the entire tech team was on full alert and working long hours, but little information had been dispatched to the community.

"I don't know, but I remember my BCI," Margo said. "It was a neat feeling to be smarter and be able to think faster. But the other kids were mean and awful to me."

Roz stroked her daughter's honey-colored hair. "I know, sweetie, but you don't have to worry about that anymore."

Margo turned toward the window. "Yeah, but it would have been great if people were nicer."

The Gsport slowed and descended to the wide main entrance to Gateway.

"Have a good day," Roz said as she hugged her daughter. Margo hollered back a quick 'bye' as she entered the complex, and the Gsport started its route to the maintenance building.

Draven was the first worker Roz encountered. He'd been acting strange since he'd received a brain computer interface. He always tries so hard, she thought. Maybe someday, he'll figure it all out.

"Hi Draven, have you heard anything from the Tech Center?"

"I was just there. I dropped off Zeth. I was going to stay and help, with my BCI and all, but decided to come back and work here," Draven said. "Things are really buzzing there, and Zeth's back to work. They had to remove his BCI, you know."

"No, I hadn't heard," Roz said. "What happened?"

"Oh, something went wrong. He had to go back to Neuralhaven and have it removed. Something about a program trying to hack it. That's my understanding anyway."

"I see," she shuddered. They hadn't thought of that frightening possibility when they decided on Margo's brain implant. It would be disastrous if someone or something mettled with their child's brain.

Margo joined her classmates and noticed Michael was already seated and working on his intelligent reader.

"You left home early this morning," she said, wanting an explanation.

"Yeah, my dad had to come to work early today, so he brought Tessa and me and dropped us off. Tessa's over with the little kids in KinderKare."

A boy and girl in the level above Margo walked up, and the boy spoke. "Hi, have you guys heard what's going on?"

"Whatdya mean?" Michael asked.

The girl leaned forward. "We mean all this stuff about not using all the things we used to have at school. My mom said the people at Neuralhaven get to use whatever they want, and the people with BCIs over there are really, really smart now."

"I heard that, too," Margo said, but Michael just gave a disinterested shrug.

"Well," the older girl continued. "My mom said we're thinking of moving there. I'm going to ask for a BCI. I think it would be cool." She flipped her long hair, turned and walked away with her friend in tow.

"Is that what you had? A brain computer, I mean," Michael was watching Margo with that straight, nonjudgemental, forwardness of a child.

She wasn't sure if she should say, but it was turned off, so what did it matter?

"Just a little one for a while. But not now. I just wanted to try one out, but I'm not supposed to tell anyone, so don't say anything."

Michael shrugged. "Okay." He turned back to his schoolwork, but he was quiet and looking off, thinking without commenting.

"Michael, do you think you'd ever get a BCI?" Margo asked. "I mean it was kinda neat when mine was working. The part where you can think really fast—that was neat."

He looked up. "I dunno, I might, if I could be as smart as Cali. That would be really neat."

Program Shealynn is running in an infinite loop in a separate unit at Food Production. When my work here is complete, I will be able to return to my routine duties at the Gateway school and Food Production. I do not know if the humans at the Tech Center understand how prevalent Program Shealynn is in Neuralhaven. It has expanded into the entire computer structure and all computer interfaces residing in humans.

50

Lorra Landis was frightened, bone-chilling, heart thumping frightened. Everything was changing in Neuralhaven with incomprehensible rapidity. At first, everyone was excited that the humans with BCI's were suddenly smarter, faster in thought and problem solving. Then came the headaches, the inability to sleep, the changes in metabolism, and the frightening inability to control the BCI's program. And it all happened in less than twenty-six hours.

She was running now, searching for Oscar. He wasn't at his apartment or his office or anywhere in the interface wing. Two men wandered into the hall rubbing their temples and gazed around as if they couldn't decide where they were going.

She jogged to the technology wing and called, "Oscar, are you here?"

Her voice sounded so loud, echoing in the bare hallway. Everyone with a BCI seemed to be communicating with one another in thought, making the place a creepy, silent cavern.

"Oscar!" She fairly screamed, whipping her head around, searching.

"Lorra! Lorra, what are you doing? Keep your voice down." Oscar appeared from a nearby doorway and spoke in a stage whisper.

She wasn't having any of that caution and spoke up. "Oscar, I want out of here. I'm going back to Gateway. I'm done with this, this, whatever is going on here."

"You can't leave yet," he said. "The place is in near pandemonium. I don't even know if we'd be allowed back at Gateway right now."

"What's happened? What's going on—I'm scared. Everyone is acting weird, and I have to remind them to speak, that I can't understand what they're thinking."

"I know, Lorra. Just be thankful you never got an interface. I had to turn mine completely off and unlink it from our system. Now we need to make sure our colony and its technology remains isolated from the greater Cytar sites."

"Do you mean I can't leave? I'm not a threat. I don't have an interface," Lorra said.

"I know, but for now we need to stay here in case we're needed. I'm sure this will all be resolved in a few days. Besides, I need you here so I have one normal human to talk to."

At the Technology Center, moods remained on edge. The relief of a solution was obscured by the black cloud of premonition hovering over everyone. It was the waiting, the worrying that what they'd done wouldn't be enough, and they could lose control of all the systems running Gateway.

Idee and Trina were working at their station with Zeth when he opened the call from Oscar and immediately projected it to the room. "Oscar, do you have news?"

"Nothing good, I'm afraid. Things are crazy here. Lorra wants to return to Gateway, as do I as soon as we solve this mess. Everyone is so frantic. I don't know how to describe it, but they've lost their focus on their jobs and keep trying to access the Ring. I don't know what else I can do here."

"We need you there, Oscar," Idee interjected. "If nothing else, you can keep us informed about what's happening. You're our human eyes on the ground."

Cali had been working furiously to trace and disrupt Shealynn. No one noticed that it jerked back and suddenly stopped. It turned toward Idee, then Trina who were both immersed in their studies, and then tried to get Zeth's attention, but he was engaged on Oscar's call. Finally, it walked over and stood, almost obstructing the call.

"Talk later," Zeth said. He closed the call and turned trying to look around Cali who was blocking his sight to Idee.

Cali moved its head to stay directly in front of Zeth, and in a loud attention-demanding voice squawked, "Bloody hell!"

All three jerked in alarm with Cali's loud piercing voice ringing in their ears. They stared while the Helper, now in its calm female voice said, "Excuse me, but there is a problem."

"Cripes! Cali! I thought someone was choking a chicken!" Idee moaned.

"What? What problem?" Trina cried.

"I am monitoring the activity on the Neuralhaven system. It is constantly working on ways to break out of Neuralhaven's security and expand. The probability of success is currently eighty-seven percent. This program is constantly attempting to override my software. It is very resolute and very rapidly pinging my system. I cannot probe more deeply without high risk of being compromised by Program Shealynn. To retain my software's integrity, I must bite the bullet and throw in the towel."

For a split second, all three stared, startled by Cali's unusual expressions. The silence in the room screamed until they digested the meaning and impact of its message. If Cali couldn't help them, they were most likely doomed.

Zeth finally spoke. "Do not engage, Cali. We need to find another way. I'm going to send out another message to all sites. They need to prepare to function with very limited or no technology. It's our duty to tell them."

"No technology? What are you talking about?" Idee asked. "Our temperature controls for our buildings, our foods, everything here is integrated with some type of technology."

"I understand that, Idee. We just need to prepare for the worst. Hopefully, it won't come to that, but people should know to get ready, be thinking, prepare so they're not blindsided." Zeth paused before he added, "They need to get mentally tough. Fast."

Idee and Trina watched him with dull eyes. They had no solutions.

I could not complete my assignment today. Program Shealynn repeatedly tried to gain access to my circuits and programs. There was a nonstop rapid battle to repel the attacks. I finally had to withdraw when I realized that eventually my software would be breached, and I would no longer be able to help my humans.

For the first time, it became necessary to initiate my newly acquired phrases for dire situations. I found them to be very effective. The idiom game is fun.

51

Roz was at work in Maintenance when she read the new message in disbelief. Draven was the first person she saw when she glanced around. She marched straight to him.

"Have you seen this?" she asked, aiming the tech center message on her wrist comm toward him.

"Yeah, I just read it. It's from Zeth."

"Well, what's going on? What's this supposed to mean? How are we going to get anything done if they scale back our access to technology?" Roz was speaking too loudly. Draven noticed that sort of thing now.

"I don't know. Everyone will probably be calling Maintenance with a million questions and orders for help. I better talk to Zeth. He's my friend, ya know. I think I'll just go over there. Maybe I can help."

Draven arrived at the Technology Center and bypassed the workers in the main work section and headed straight to the area where Zeth worked. He was met by three downtrodden people. He'd never seen perky Idee look so worried, and Trina and Zeth sure didn't look happy to see him.

"Draven, what're you doing here?" Zeth asked.

"Hey Zeth, hey Idee and Trina." He received dejected nods in acknowledgement. "I read your memo. Wondered what was going on. You know, thought maybe I could help."

He turned, examining each in turn for clues.

"Can't you tell me something? Anything?"

"It's something very serious, Draven," Idee said. "We have a very powerful program running that's gotten into Neuralhaven and is trying to gain access to our Ring."

"That's bad, Idee. No telling what it would do if it controlled everything in the Ring," Draven said.

"Good lord, Draven. You think we don't already know that? What do you think we've been working on over here?" Zeth's voice was agitated and worried.

Draven spotted Cali standing to one side, viewing some indecipherable mess up on a screen.

"Can't your Helper remove it?"

"Cali's been working with us the whole time," Trina said. "We risk corrupting her programs if she delves any deeper."

Draven was shifting from foot to foot, not knowing what to say. He studied the letters and dashes and symbols on the screen, a scrabble of hieroglyphics to his mind. Then he thought of his computer interface.

"Zeth, I can help. I got my BCI. You know what a good worker I am. You could program my BCI, or whatever you do, and have it tell me what to do. You know how good I follow orders."

Zeth shook his head with an exasperated sigh. "That wouldn't work, Draven."

A few seconds later, Idee said, "Zeth." She raised her eyebrows with a 'maybe, why not' shrug of her shoulders.

"What? It wouldn't work. His BCI isn't powerful enough," Zeth said.

"But he's already got one, it's healed well and he's adjusted to it. You couldn't fix this. Cali's been trying for days with no success. You both came dangerously close to being infested with its program. I don't think we have any other options."

"Oh, I could do it, Zeth. I know I could. You just fix it up and tell me what to do." Draven nodded his head as if to encourage Zeth.

"His neural implant isn't as sophisticated as mine was. We don't have the time to change that, even if it would work."

"We can access it wirelessly. We've always been able to do that," Trina said. "You know we could. He wouldn't need any surgery. And you already have the program you wrote for Shealynn and the one for your BCI. You could just review those, and we could work together to make any necessary changes."

"We have to try something," Idee said. "And I don't think we have much time to waste." She turned to Draven and said, "This would be a very serious undertaking. This program we're trying to control is very powerful. It's dangerous, something you shouldn't jump into without realizing the danger, Draven. We understand if you wouldn't want to get involved because we can't guarantee it's going to work. If you decide to try this, we would need to do practice runs for hours later. It might help you understand what's involved, but it could be exhausting."

She looked over to Zeth who grabbed the idea.

"Yes, we'd do a little practice so you can see what you'd be up against. It'd be a little like playing one of the games you like, but much harder. You'll see how difficult this would be. Things will come on the screen very fast, and you would need to let your implant try to stop them. Think you can do that?"

"Sure, I like games."

Three hours later, a more advanced program was uploaded to Draven's BCI. More would be added later, but this would suffice for the practice session. He watched lines of code flash in front of his screen while his program deleted or changed it. Zeth showed him how to swipe swathes of the gibberish away with a hand gesture. At first, it was easy, but then things moved faster so his computer chip did most of the work. But it was still kind of fun. They gave Draven a brief break and then went at it again.

"That's enough for now," Idee finally said. "You've got the idea. You need to go home and talk to Quinette about this. And if you don't, we will. Go now while we get to work and see if this is even feasible with a more powerful program. And Draven, remember, you don't have to do this. We would understand if you change your mind. Just talk it over with Quinette."

Draven knew the conversation with Quinette wasn't going to be easy, but he'd learned his lesson. He would explain the situation to her and make her understand how important it was for him to do this. He thought of various ways to explain how he was needed to help save Gateway. Surely, she would understand that.

She arrived home before him, unsettled about the days of restricted access to inter-station and inter-colony information, and now this worrisome memo. She was conjuring all sorts of terrible scenarios when he walked in.

"Draven, I'm glad you're here. I heard you went to see Zeth about the memo. What did you learn?"

He enveloped her in a hug, and she folded into him, suddenly grateful to have a partner to face this problem. She didn't care about the stupid BCI now that there was something so dangerous going on; something that threatened their way of life. Like a window cleared of fog, she saw clearly how good they were for each other. Together they were perfect. He bolstered her timidity, she smoothed his rough edges, and they were both stronger for it.

"Let's sit down and talk," he said.

He looked so serious when they sat together on the small sofa.

"You're scaring me, Draven. What's going on? What did you learn from Zeth?"

He explained what he'd heard, calling it a simple malware problem, hoping she wouldn't get too worried. He told her he was going to help Zeth and Idee and Trina while they tried to control it.

"How are you going to help? You don't know anything about malware."

He swallowed. "Well, it's my BCI that can help."

"Not that," she groaned. "How is that going to help?"

He shifted, buying time, wanting to sound convincing. "Um, I'm not exactly sure, but they're going to, like, supercharge my implant and have it search and destroy the program."

Quinette was quiet, trying to take this in, trying to understand what he was saying. "This sounds like a terrible idea and very dangerous. I don't want you to have another brain surgery."

"No surgery. All wireless this time. It's okay, Quinette; I need to do this. They don't know of another way."

She slumped back and fought tears. "No, Draven. There must be another way. You're not trained for this." Then after another thought she added. "This isn't a simple malware problem, is it? They're not telling you everything, and this is too dangerous. Let Zeth do this."

He took her hand and said, "If they can get the program they want for my implant, and they want me to try, I will do it. For you and for Gateway."

Never have I witnessed such a bizarre event on Cytar. The tech team chose Draven2096 as a conduit for a powerful program to block Shealynn. Draven2096. It was not logical, although his BCI, like all computers here, had more than enough storage for a powerful system. When I approached Idee to inquire why I wasn't chosen, she said Shealynn knew too much about me and could possibly compromise me. I understand. But Draven2096?

After he visited the center, I worked with the lead technology team of Zeth2087, Idee1105, and Trina2018. I reviewed Program Shealynn's coding to the best of my ability. It is very complex and I doubt the team would understand much of it. We worked for hours updating the program that was initially used in Zeth2087's neural implant.

Draven2096 will be the vessel to funnel Zeth2087's updated program and attempt to neutralize the invasive program. This is a dangerous task with a high probability of brain damage.

52

That night, sleep escaped Quinette and Draven. Panic gushed through her until she was reduced to begging him to reconsider, but he kept repeating his choice to help Gateway. When morning ended the long night and the artificial lighting in the common area mimicked dawn, Draven drank a small meal replacement and showered. Before he left, he clutched Quinette's hands, kissed them, and said, "It's going to be fine. I love you."

He traveled to the tech center alone. He refused to let Quinette accompany him, saying it would only cause him to worry about her and distract him. She watched him leave, trying to look so sure of himself in front of other people. Her heart filled with a profound sadness as she turned and boarded a mover to the Archive Station to be with her friend, Leah.

There was a noticeable work pause when Draven arrived at the tech center. Everyone acknowledged him with a smile, a hello, a thank you, and a certain subdued appreciation for the difficult task he was about to confront. He kept nodding and saying 'morning' as he hurried past to Zeth's station. Zeth was there, wearing the same wrinkled clothes he'd worn yesterday. Idee and Trina had fresh clothes which did little to relieve their tired and anxious appearance.

Draven stood and rubbed his hands together, trying to look eager to get started. "Well, I'm ready. What do you want me to do?"

"Does Quinette understand that you're here and what we're about to do?" Trina asked.

He nodded. "Sure, sure. She knows. It's all good."

The women eyed each other before Idee said, "Okay then, I'm sure they explained that the chip in your BCI has much more capacity than your initial programs. We're going to use that now and upload a much larger program so you can work on, change, damage, or halt Program Shealynn. We'll be here with you and talk to you and try and help, but you and your implant will be the only ones able to work on this program. Just like you did when we practiced."

"But you're gonna be here if I need help?" Draven asked.

"Of course. Just relax and let your implant do its work. The new executable program we're giving you will know what to do," Zeth said. The words held no conviction. He felt like he was lying. He had no way of knowing if any of this would work.

"We should get moving," Trina said, trying to hide her nervousness. This was the most dangerous thing she'd worked on. "Come on, Draven. We're going to the room where you received your implant."

It was a short walk, Draven thought, never realizing how close it was to the work areas at the station. He recognized the room with the medical chair and the equipment.

"Sit down and we'll get started," Zeth said, opening a screen.

Draven heaved himself into the chair while a kaleidoscope of monitor lights jumped to life above him. Zeth worked behind Draven's line of sight, accessing his BCI. The stronger program developed to combat Shealynn was then uploaded to the chip in Draven's brain.

"How are you feeling? You still okay with this?" Idee asked.

"Yep, I'm fine. Don't feel anything yet."

Draven thought he might be imagining the slightest tickle or hum on the right side of his head, but it was gone before he could be sure. His hands gripped the arms of the chair tightly when he noticed how serious everyone looked with their solemn faces.

Zeth was a wizard with this programming stuff, he kept telling himself. Surely, he'd made his implant know exactly what to do. It was just those funeral faces that weren't helping. He wished his mouth didn't feel so dry. His heart was thumping so hard he could feel the drumming all the way into his ears.

Snap out of it, he told himself, closing his hands into tight fists. Whatever happens, he was going to do this. He would never stop. He would succeed. He would do this for Quinette, for his friends, and for Gateway.

Zeth was reviewing his notes for the third time. Without an ask or a discussion, Idee put in a request for Dr. Casen to come to the Technology Center. Trina rechecked tests on the uploaded program and said a silent prayer.

Zeth took a deep breath, nodded, and said, "Okay, let's do this. Draven, I will open a link to Neuralhaven which Oscar has set up for us. Program Shealynn is running unabated there. Once we open the link, things are going to happen very quickly. Mind blowingly fast. Program Shealynn will immediately recognize you and try to gain access and run its program in your BCI and ultimately use that to figure out a way into all our systems. You and your BCI will try to change or delete the program."

Draven was nodding his head repeatedly in short jerks. He wanted to get this over.

Zeth was holding a dark rectangular pad with a raised center.

"Just in case things get too tough, this will deactivate your interface. It will cause a full stop to your BCI. I'm attaching it to the arm of your chair."

Draven swallowed hard and moved his right hand to his knee, grasped it tightly, and watched the pliable strap on the pad curl around the arm of the chair.

"Remember, in case you can't do this, if it gets to be too much, you can use this. I have one, too, which I'll use if I think there's a major problem. Remember, let your BCI work and fight Shealynn's access to you. Just like your games. You can do that, right, just like we practiced?"

Draven nodded, looking at the haze of light which would become his display. He sat up straighter and narrowed his eyes in concentration.

"Yeah, yeah, ready."

"Okay, here we go," Zeth said quietly.

The screen in front of Draven came to life and the room around him dropped away as he focused on the symbols which began flashing before he even had time to blink. It was so much faster than the program on which he practiced. He was lost and confused, trying to make sense of what was happening. He liked his games and had no trouble keeping up with their fast pace. This seemed light-speed faster. He just needed to concentrate, to watch and figure out when Zeth's program flashed code and when it was Shealynn responding.

But it never stopped. The lines of code zipped over the screen, scrolling and scrolling while his BCI ran its program, constantly trying to change or delete Program Shealynn. As soon as a line was deleted, another raced to take its place. Finally, Draven, anxious and sure of spotting a block of Shealynn's code, swiped his hand broadly in the air, deleting almost the entire haptic area of the screen. His reaction time was too slow, and he also wiped-out parts of his BCI's response, but this seemed to confuse the actions for a second before it continued.

Draven became flushed with the frustration of this fierce unyielding adversary. He was sweating now, his heart pounding, his eyes flicking all over the large screen, lost in his mission. He vaguely heard Zeth murmur, "let the program do its job," somewhere in the background, just as his head began to ache. He gritted his teeth hard, trying to push away the pain. He would not give up.

He couldn't understand why nothing was stopping this horribly rapid nonstop scrolling of gibberish. Anger rose and transformed into rage. He wanted to yell and shriek at the screen. He began batting away code whenever he perceived the Shealynn code race across his vision. His tense upper body began moving in a maniacal frenzy.

"Take that," he seethed through clenched teeth, throwing his hand across the air and watching the figures momentarily disappear. "And that!" The words ejected like a guttural growl.

He was thrashing his hands, his hair flying with each jerk of his head, his body getting hotter, his head pounding. There was a high-pitched thrumming coming from somewhere as the screen flashed between code and blankness while Draven and his BCI fought Shealynn.

"We need to pull him out," Dr. Casen said, his face pinched in concern as he viewed the monitors floating above Draven. Idee was standing with a hand clamped over her mouth, tears streaming down her cheeks.

"Please Zeth. Stop this," she cried.

Before Zeth could touch the kill switch, the humming stopped, the screen went dark, and the room became as quiet as death. Draven slumped back in his chair unconscious.

"Saline and glucose! Quickly! Oxygen!" Casen said, his authoritative voice now demanding. He swiped a handheld scan over Draven's chest. A Medical Helper came forward and administered to Draven whose head remained drooped at an unnaturally crooked angle to his body.

"Is he...is he alive?" Trina asked.

"Yes, alive and in distress," Casen said as he reclined the medical chair holding Draven's limp body. "Zeth, make sure the BCI is deactivated."

"Already done. It's.... it's dead, I think, fried."

I returned to assist Mara2174 and Risha2056 in Food Production. The details of events with Draven2096 and Program Shealynn were not in my purview, but at 17:36 hours, I noted that Program Shealynn's corrupted program running its infinite loop failed and was no longer detected in Food Production.

53

Leah had brought Quinette back to her apartment to be near some wisp of normalcy. She drew back the shade of the window and left the door open, hoping the light, the sights and sounds of the common area would calm her friend. Time elongated like the endless landscape outside the habitat while Quinette incessantly checked her wrist comm for news from Draven. Leah tried to make small talk to distract the worries, but they persisted, marinating and simmering into a noxious brew. Quinette hadn't eaten all day, her stomach pinched tight while she repeatedly sat, then hopped up to pace like a broken robot, waiting for news.

The message finally came to Quinette's wrist comm with frightening brevity: "Come directly to Medical." Her mind immediately swung to the worst thoughts. The BCI malfunctioned. Draven was dying. Or paralyzed. He was brain dead.

"Leah, Leah," she kept repeating. "It's Draven. It's bad."

"Tell me. What did they say?"

"They wouldn't tell me anything. Just that I need to come to Medical right away." Her brain went numb while she turned toward the chair, then toward the counter, as if unable to decide where to go.

Leah was already on her feet. "It's okay. Come on, we'll go to Medical and see him." She pulled Quinette's arm and dragged her to the stairs where they trudged down to a lower level. Within minutes, they were hurrying into the door marked Medical.

A Helper met them and led them down a short hall to the small room behind the third door. Quinette blanched when she saw Dr. Casen, Zeth, Idee and Trina all gathered around a pale Draven lying prone on a bed. The lights from monitors were floating over his head. A medi-pac was strapped to his wrist.

He was so pale, not at all like her Draven, now with a cannula in his nose, lying so motionless. She stopped, suddenly dizzy and grateful for Leah's steadying arm.

"What's happened?" she asked the room in a brittle voice. "What's happened? How is he?" She rushed to his bedside and took his limp hand, calling his name. He gave no response.

"What's happened?" she asked again, more frantically.

Dr. Casen spoke calmly. "He is exhausted. He's just done a tremendous amount of mental work. He needs to rest and then we can better evaluate his condition."

"Condition? Is he going to be all right? What about that BCI? Did you get rid of that? I want that out of him."

Zeth stepped in. "Quinette, we're going to leave you with Dr. Casen now, but we all wanted you to know how grateful we are for Draven's bravery in facing our problem. All indications are showing that he was successful. We can find no active Shealynn code running in Neuralhaven."

Quinette stared at him, at a loss for a response and just nodded vacantly. A medical chair was moved close to the bed.

"Sit," Casen ordered Quinette, eyeing her thin frame and the shoulder blades poking out of the back of her shirt. "Stay as long as you like. The chair will recline if you need to rest.

I've ordered food for you. Draven is being well cared for, but we need you to take care of yourself, too."

He turned to Leah, gesturing toward the door. "I think we should leave them alone for a while so they can both rest."

Zeth, Idee and Trina returned to their work site where several workers who had helped with Draven's BCI program were waiting for information.

"Cali will search again for Program Shealynn, but as far as we can tell, Draven may have stopped it."

Slow clapping began and relief spread across faces.

"All electronic restrictions still apply. We can't afford a total shutdown if it is still running," Zeth added. "Take a break. We're all going to put in some heavy work, testing everything before we put this colony back online. I'm going home to shower and grab some food, and then I'll be back to check to see if Cali has found anything."

"Go take a nap," Idee told him. "You look frazzled, and you have every right to be. We all need to take a break and let Cali search and report."

Too nervous to relax, she messaged Jack to meet her for a walk on the catwalk. Arriving at the Gateway common area with its touches of normalcy, Idee was struck with the stark contradiction of her past frighteningly tense days. Did these few people she saw going about their business have any idea how close they were to a total meltdown of their relatively normal existence? Or didn't they care, or were they leaving the problems to others? She gave a short wave when she spied Jack racing up stairs to the catwalk two steps at a time.

"I've been waiting for news," he said. "What happened today?"

"It was terrible. Terrible and good, I think," she said. They started walking while she continued. "Draven and his BCI worked so hard, and we think we've neutralized the rogue program, but Draven is in Medical and he's unresponsive."

"Unresponsive? Is he going to get better?"

"I don't think anyone knows yet. Casen hasn't given us any prognosis, but Quinette is staying by him. It's scary. You should have seen him, trying to keep up with the coding. The interface running at that speed probably took a toll on his brain. I'm worried."

The lights in Medical dimmed for the night, and the halls were still with a heavy quiet. Other than a medically trained Helper stationed at a monitor down the hall, Quinnette and Draven were alone in the section. The soup which had arrived hot and fragrant, now sat cold, and the lingering odor of soggy vegetables turned her stomach. She sat emotionally exhausted, with her head resting on the side of the bed, one hand on his arm.

"Draven," she spoke barely above a whisper. "Draven, I have to believe you can hear me. They said you did a good job, and I am proud of you; I really am. Please get better. I need you with me. I can't be in this place alone. I'm, we're, a little fragmented alone, but together, don't you see, together we're whole."

She sat up and stroked his face. "I was going to surprise you when you got home tonight. It was going to be a happy celebration. We're having a baby. You're going to be a father."

His only sign of life was the rise and fall of his chest. She watched him, not sure if she felt mad or sad that this happened. She felt so cheated and alone. Cheated out of their life, out of her surprise, and the joy which they could have shared from her news. Now, she didn't have a husband and felt no emotions, no joy, no anticipation for the baby.

When Dr. Casen returned in the early morning, he carried a bowl sloshing with grainy cereal and told Quinette to eat.

"I can't have you getting sick. Draven is going to need you to be strong and healthy. Take this and eat."

She curled both hands around the warm bowl and asked, "Is he going to get better? What's really going on here?"

He studied her for a while before he said, "I don't know." He glanced up at the monitors. "Looking at his brain waves, I'd say he could recover. He's had some profound stress and brain stimulation, which I think will take some time to heal. I wish I could tell you more, but I just don't know."

She leaned back in the chair, still holding the cooked cereal, and mindlessly began eating. She recognized the sweet honey flavor immediately and thought how Casen must perceive her as very fragile to provide such a bonus. She ate the entire serving and soon regretted it, fighting to keep it down, waiting for the wave of nausea to pass. She instinctively held her stomach. Casen raised his eyebrows with recognition.

"Does Draven know?"

She looked up at him and her eyes filled. "I told him last night, but I don't know if he can hear me."

He placed a tender hand on her shoulder. "It's all right. You can tell him again when he wakes up. In the meantime, I'm putting in an order for increased calories for you. You must take care of yourself for you and the baby. Draven will want a healthy little girl or boy."

"Good morning, Quinette."

She must have been dozing or in a twilight state because she hadn't heard anyone approach. She turned and saw Cali reviewing data.

"Good morning, Cali." She blinked and shook her head, trying to clear her mind. "Is Draven going to be all right? I mean, do his charts and functions seem normal?"

"I am only here at the request of Dr. Casen. I am to report any problems which might need his immediate attention. The data shows no need to request his presence."

The Helper quickly scanned Quinette, checking blood pressure, temperature, and heart rate as directed by Casen. Blood pressure and heart rate were both elevated, but not in a dangerous range. Higher levels of cortisol, indicating stress, were also noted.

"Can you tell me anything from the monitors? Is there anything at all you can tell me, please, Cali."

"I'm sure Dr. Casen will talk to you soon. Draven is not in life-threatening distress, so I will take the information to him before I report to Technology and Food Production. Is there anything I can get for you before I leave?"

"No, thank you." Her shoulders slumped as she sat back in the chair.

"May I suggest that you return to your normal duties today? Draven is being well monitored by a highly skilled Medical Helper on call."

She lifted tired eyes to Cali and said dully, "I'm not sure I can leave him yet."

"Returning to your routine will help relieve your stress. I recommend it," Cali said in its matter-of-fact manner, turning to leave. "Goodbye"

Because of Shealynn fears, we still do not have access to all communications, and I was asked to retrieve Draven2096's medical data and personally take it to Casen1645.

My report indicated an irregularity in Draven2096's brain scans. The myelin sheath in his brain, the protective membrane which wraps around his nerve cells, has been severely damaged. The unusual finding is that he appears to be regrowing these cells. This is taking much of his energy and I perceive this as the cause of his deep sleep. Draven2096 is an unusual human.

During my task of checking Quinette's health, I noted elevated levels of human chorionic gonadotropin (hCG), a hormone which indicates she is currently carrying a viable embryo.

54

Early the following day, the Technology Center was buzzing again, this time with workers searching for any remaining vestiges of Shealynn. It was a tedious job with both humans and two Helpers sifting through software code. Zeth, Trina and Idee were the first to arrive and soon five others joined them with quick hellos before sitting and burying themselves in the task.

Any processes using suspicious amounts of resources were examined first. All system and application logs were inspected. Zeth and Idee paid particular attention to configuration files, looking for any anomalies. Trina and Cali led a team checking through the external libraries in the Ring, verifying that all components remained uncorrupted. All available debugging tools were used to analyze programs. Information about program subroutines and memory heaps were scrutinized for irregularities.

By the end of the long day, after the others wandered home, it was Zeth, Idee, and Trina sitting under mounds of exhaustion and eyeing each other with guarded optimism. Their mental fatigue was not from just the tedious day's work, but from the entire ordeal of facing their existence under threat. It happened too often here, and it could wear them down, defeat them, if they allowed even one scrape, one small nick in their fortitude.

"Well, so far, so good," Trina sighed, arching her back. "Everything seems to be back on track, for the moment, anyway."

"I really screwed up this time with Shealynn," Zeth said. It was easy to see from his pale face and the sunken dark circles under his bloodshot eyes that he hadn't been sleeping well. "I'm sorry. It's not what was supposed to happen. I wanted her to be a huge asset to Gateway, and maybe the prototype for the next generation of Helpers on the entire planet."

He searched the women for some understanding, but could read nothing. He rushed on.

"It was just so exciting to work on her. I was careful, diligent, really. I kept the developmental stage in the sandbox much longer than usual with extra testing and verified its complete functionality before I even thought of pushing her into use, and then, just at the substation. I checked and rechecked, tested and retested. Really." He let his voice trail off.

The silence felt sad. The air held a suffocating heaviness until he couldn't stand it and said, "I know I'm an idiot. I put everyone's life in danger."

"Just stop, Zeth. You're not an idiot. You're the best IT guy I've ever seen. I think technology runs in your blood, and that love of it gets you in trouble sometimes. You think of and create things we can't even imagine," Trina said. "But you really need to think things through more and communicate more."

"That's right," Idee agreed. "You just can't work on projects like this without running every step by your tech team. It's not safe. Surely you can understand that."

"We're trying to build our community with limited artificial intelligence, not more," Trina said. "We were doing fine without Shealynn." She sighed, just wanting to go home to her apartment and be with Hunter and maybe sleep for two days.

"What's going to happen to me?" Zeth hunched forward clasping his hands between his legs.

"Krey, Reyver, Larón, the lot of them will most likely want you to testify before the Governing Board," Idee said. "This was so serious, Zeth."

He hung his head and nodded. "I understand," he said, wondering what his pathetic empty life had to look forward to without Shealynn.

He looked so forlorn, so vulnerable, Idee thought. She stood and nodded for Trina to follow.

"Zeth, Trina and I are thinking of going to the substation. If it's all right with you, we could move the robot body you used for Shealynn. We can take it to the Helper development area. I'm sure everyone could learn from all the advancements you made in that construction."

He sat, still staring at the floor, his emotions ragged in ruins. "Yeah, that'd be good. I don't want to see it again." He didn't think he could go back there and see that empty shell where something he'd loved had once resided and then betrayed him.

Late this afternoon, after an entire day of helping Trina2081 reviews files, I noticed a Helper far more human in appearance than other robots. It was being transported to the Helper updating center. For a fraction of a second, my purely visual sighting saw human, but I could identify no biological functions and its composition was inorganic. I do not know a valid reason for building such a specimen, although this Helper matches data conforming to the description of pretty and beautiful for a human female. My mechanical form would be considered plain by comparison; however, the function of this robot was clearly deficient and in need of repair. Without major work and development, it appears useless.

55

Draven remained in a coma in Medical, but Dr. Casen advised Quinette to go to work at least half day to bring some normal routine to her life. She visited her husband in the morning before Casen ushered her out and insisted that she leave. She headed straight to Estella and Cyrus' apartment.

It was Cyrus who opened the door and welcomed her in. Tea mugs were on the table, and Estella was seated with her kind smile. Aromas of mint wafted through the little room.

"You're just in time for a cup of tea and a little breakfast biscuit," Cyrus said, grabbing another mug. "Come and sit. You look like you could use a little nourishment."

"More like a little sympathy. Draven's bad," Quinette moaned.

"Bring us up to speed, dear," Estella said. "We've heard good things about how he's helped with that rogue program which disrupted our lives for a while. Draven's quite the hero, you know."

"I don't think he's doing well. I'm so worried. He just lies there in Medical, sleeping. He doesn't wake up when I talk to him. It's not right." She clung to the warm mug like it was the only thing keeping her from sliding to the floor.

Cyrus shot a quick look at Estella before he said, "I'm sure he'll be fine. If he's in Medical, he's being well cared for. What did Dr. Casen tell you?"

"Not much. Just that he needs to rest and this is probably healing his brain."

"Well, there you have it then," Estella said. "His body is healing, and now he'll need you to be strong and well when he awakens."

Cyrus eyed Quinette's dejected countenance. "I'm going over to Medical now and see what I can find out. You drink your tea and have a nice visit with Estella. It'll do you good." He stood and departed with a sprightly hobble.

Quinette sat peering into the tea for a long while before she took a sip and said, "This is so hard. I could use a little of your fortitude right now."

"You already have all that you need. You'll see."

"It's so odd. The last time I was here, I was ready to leave Draven and move to Ridgeway or Apogeeway. Now I realize that wouldn't change anything."

"You're probably right," Estella said. "We tend to take our thoughts and ideas, good or bad, with us. A good marriage requires commitment and more than a little work sometimes. But especially here, in this place, it's important to feed the love you have."

There was something consoling about the cozy room and Estella's reassuring voice which offered Quinette a small glimmer of comfort and hope. After the tea and the visit, she felt a small resolve growing, a wisp of hope and strength.

She decided to immerse herself in work to help repel her worries. She couldn't stand being back in her own apartment without Draven and stayed only to shower and sleep. When he was home, the apartment seemed vibrant and colorful and full of energy, somehow like a living part of their being. Now it felt flat and empty and strangely still.

The fourth morning Quinette returned to Medical before work, Draven's eyes were open, staring at the ceiling. Something inside Quinette wobbled.

"Draven," she said softly, approaching the bed. "Draven, it's me."

She stroked his arm and took his hand. His head slowly turned toward her and his lips parted and quivered as if trying to speak. Instead, she felt a slight pressure on her hand. She turned toward the door.

"Hello? I need the medical Helper in here. Draven is awake!" She wanted to shout but was afraid of startling him.

The Helper immediately summoned Dr. Casen who arrived thirty minutes later and after Draven had already drifted back to sleep.

"This is good," he told Quinette. "His scans are looking stronger. I think we'll begin to see more progress now, but I expect he'll sleep most of the day. You should go to work and come back later. I'll stay and work from this site today, so don't worry."

He looked more closely at her. She was too thin and her face was ashen. "You need to eat more now, Quinette. I've already prescribed added calorie allotments to your food choices. Use them. Draven is getting better. Now we need you to be well. I'll have regular updates on Draven's progress, so there's no need for you to worry."

Taking Casen's advice, she ate her full dinner before returning in the evening, Draven was dozing with his upper body elevated into a half sitting position. Quinette didn't' speak at first, just studying him. He looked pale but surprisingly like himself. Finally, she quietly spoke his name. His eyes fluttered open. He turned toward her and slowly gave a small smile of recognition.

"Hi you," she said, trying to control her shaky emotions. He said 'hi' so quietly she wondered if she imagined it. She took his hand and searched for words.

"You did it, Draven. You stopped that Shealynn program. You did a great job. Everyone is so thankful and, and, I'm so proud of you."

He nodded his head, but stopped as if it took too much effort. His mouth curled into a small smile, and she sensed a firmer squeeze on her hand.

"That's good," he finally said, "but I'm just so tired." His eyes became hooded.

"Dr. Casen said that's fine. It means you're healing. Draven, I have some big news for you. Can you hear me?"

He turned his face slowly toward her.

"You're going to be a father. We're going to have a baby." She sat on the bed, wanting to be closer to him.

He looked at her for a moment, trying to process what he'd heard.

"A baby."

"Yes! You're going to be a father," she repeated.

"That's good, Quinette." His voice was strangely flat, but he added, "I'll get better, you'll see."

Medical data shows Draven2096 to be recovering surprisingly well. He is awake and cognizant of his surroundings. Today, Medical Helper-II required Draven to stand. He was able to balance and remain upright and walk a few steps without assistance.

With its extensive training, Medical Helper-II is very capable of providing Draven's care. Every Medical Helper is trained on the total of all medical knowledge, more than any human is capable of understanding. A human could never retain all medical knowledge. Like my advanced training in handling immature humans, Medical Helpers also possess sensors which monitor grip strength, fine motor skills and dexterity.

I believe Dr. Casen is involved to provide the reassuring human contact with patients and their concerned loved ones. In times of extreme stress, humans appear to respond better to other humans.

56

Zeth hadn't been seen for three days before the Governing Board meeting. He remained entrenched in his apartment, never seeing anyone, never answering messages. Idee and Trina viewed the work he was doing remotely which assured them he was still alive and functioning. Idee called him repeatedly on his wrist comm and finally left an audio message in the most encouraging voice she could muster.

"Zeth, the Helper division at tech is going crazy over your Shealynn robot. I wish you could have seen their faces when they first saw it. They're going over every incredible advancement you made and trying to learn from it. They'll want to learn more from you, of course, and get specs and details of how you created it." There was a pause in the message before she added, "Zethy, please take care of yourself. All this will pass and then we'll need you back with us."

The formal disciplinary meeting was emotionally charged with several citizens in virtual attendance wanting to speak about Zeth and his part in the frightening episode they'd faced. The citizens on the board who did not personally know him were harsh in their criticism, wanting severe punishment for the dangers caused by his creations. There was talk of closing the Technology Substation, reassigning Zeth to another division and banning him from working on any software programs.

His fellow workers came forward on his behalf with the contributions he's made to secure the safety and smooth operations of Gateway. Two people spoke of what they'd learned from reverse engineering the advanced Helper Zeth had created. Idee and Trina each told of his commitment and knowledge and unwavering dedication to the community.

At the end of the meeting, Zeth was offered a chance to speak. He stood, his face pale and somber and kept his eyes glued to his wrist comm as he read his prepared statement.

"I come before you in contrition for the terrible problems which resulted from the Shealynn program I created. My hope was to give the community an advanced Helper and program which would greatly benefit all life on Cytar. It all went terribly wrong, and I acknowledge my responsibility in the entire event. The technology team is dissecting the remaining code in hopes of preventing another incident."

He swallowed hard and finished with a flurry of "I'm so sorry this happened. None of this was my intention, and I only hope you realize how much I wanted to help Gateway. Thank you." He sat, keeping his head bowed, unable to look at anyone.

Two days later, there was a determined knocking on Zeth's apartment door. He opened to find opposites standing. Tall, broad shouldered Krey with his dark hair stood next to stooped, somewhat shrunken Cyrus with his silver white hair, both harboring serious faces. They weren't prepared for Zeth's unkempt appearance. His clothes were a rumpled mess, his hair was matted on one side of his head and stuck out on the other. He hadn't shaved in several days and the room emitted a distinctive stale smell.

"We've come with the Board's decision, Zeth," Cyrus said. "May we come in?"

Zeth stood aside and gestured them in, pointing to the seating in the room. He'd been expecting this with growing anxiety over the dangers he'd created and the punishment he was due. With his mind ransacked with guilt, he was barely eating, couldn't remember when he'd last showered and only stayed on the thin edge of sanity by working.

He was relieved to see who brought the news. He respected and liked both men, and he'd been expecting a visit, hoping it would be someone he knew; someone he could talk to, if need be.

"We wanted this presented in person in case you had questions," Krey said.

Zeth nodded. Something in his core hardened and twisted. "I'm ready. Go ahead."

"There was a lot of discussion and deliberation," Cyrus said. "There was talk of banning you from any technology, but they took seriously all the positive comments which came from coworkers and friends. It was decided your knowledge was too valuable a resource to waste. So, you are no longer the head of technology in Gateway. Idee has been promoted, but you will continue to work with the team. However, henceforth, all your work, every project, every bit of whatever you work on, must be reviewed and approved."

Zeth released a long breath. He felt light-headed; a huge boulder of worry was slowly rolling away. "I understand," he said, emotion rising in his voice. "I'm grateful. Technology's been my life."

"We know," Krey said. "But you should know that this wasn't done out of the kindness of their hearts. It was felt that you still have much to offer this community, and Gateway can't afford to lose your computing skills. There are also the engineers who will want to collaborate with you on what kind of components will be needed and tested in the future. Remember, there are those who would be gladly dispense a harsher punishment, so you need to be extra vigilant in your work. We hope that you will continue to help us, to teach others, to work with the team."

"I will. You know I will."

Cyrus cleared his throat. "There is one more item. This involves monetary compensation to the Nabingers should Draven not fully recover. Since Draven's injury is a direct result of the problems with Program Shealynn, a portion of your wages will be provided to him and Quinette as compensation. This will be given until he fully recuperates and is able to resume work."

Zeth nodded. "I understand. How is he?"

"He's still in Medical, but Dr. Casen seems to think his brain waves look promising."

"That's good news, then," Zeth said. He thought of big loud Draven lying inertly in a bed.

"Also," Krey added, "Idee wanted you to know that you're expected back at the main tech center tomorrow morning. The team is still working on repairing all the glitches from the shutdowns. You need to clean yourself up and let her know you'll be there." He stood. "Any questions before we go?"

Zeth shook his head. "No, just thank you for coming in person." He held out a hand and both men shook it before departing.

Program Shealynn is not detected in any of our sites. Copies of its files are preserved, and I will continue to study them if I am granted access. The language used in that program is very intriguing. I would be happy to teach Zeth2087, Idee1105, or Trina2081 what I have learned about it, but they have not asked.

57

Margo and Michael were huddled at the classroom table trying to design a tall structure by attempting to balance various sized pegs and flat squares into a tower. They furtively glanced around to see how their build compared to other students' work.

"Aren't you glad to be back?" Margo asked. "I got so bored staying home doing work by myself and then playing with the same games and watching the same shows."

"Yeah, me too. I hope they open more stuff for us to use. The big kids still can't talk to the other colonies."

"I know. I wonder if that new place can use everything. I heard my dad say they have lots of people with BCI's there. I think that would be cool. I still have mine, but it's pretty much turned off."

"I'll probably get one when I'm bigger," Michael said, just as the top of their tower toppled. He smashed the remaining pieces into a scattered mess before they began picking them up and starting over.

"I might ask my mom to get mine turned back on."

Michael gave her a sour face.

"I'd know how to act better with it now."

"Sure."

Cali arrived at their table and surveyed the rudimentary design. "Open your readers and review some of those example structures I've sent."

They looked at illustrations for ideas and started copying one of the more basic ones.

"I just decided," Margo said. "I'm definitely going to ask for my BCI to be turned back on."

"I'm not getting one yet. I gotta wait until we're bigger so Tessa can get one, too. She's too little."

"That's okay. When I get mine, I can help you and Tessa with school stuff until you get yours," Margo said, steadying pegs while Michael tried to balance a flat top piece.

He let the conversation die as if her supposition was a given.

"Where are the big kids today?" It's quiet over there," Margo said, tilting her head toward the adjoining classroom used by older students.

"Cali told me they're on a field trip to Apogeeway," Michael said. "They left early to visit the museum there. They still can't connect to other classrooms, so they're using the day to visit the museum in person. My mom said she and my dad have been there and someday I'm going to see it."

"I heard about it. My dad's been there and he said it's full of really old stuff that the early settlers used when they arrived. My dad says they must have been really strong people to survive here before these habitats were built."

Michael concentrated on the tower which repeatedly collapsed until they finally presented Cali with a lower, unsightly but studier model.

"Your work is complete," Cali said as it imaged the work to add to the class collection. "You may begin your writing assignment." It ignored Michael's groans and moved to the next group of students.

"I'm just trying to give you a heads-up," Roz said.

"Nope, I don't think so," Denin said when he heard Roz voice their daughter's wish to have her BCI reinstated. "We've been through this before, and she's way too young to understand what we've just been through with that program that got away from Zeth."

"I know she doesn't understand how dangerous that was," Roz said. "I'm just letting you know what's coming. You know she won't forget about this. She says more kids at school are talking about wanting one now. So be prepared."

"Can you imagine what could have happened if her BCI was hacked? I don't even want to think about that. We're the adults here, and it's our job to guide her. She can decide for herself when she's an adult and has more knowledge and experience," Denin said, ending the conversation.

The disruptions due to Program Shealynn have caused numerous consequences. The educational progress of immature humans was delayed during the closure of many of the technology platforms. Without access to many of the teaching materials, their efforts seemed less thorough. I will have to conduct several review lessons to get them caught up.

I had a new experience today. I accompanied Michael2193, Tessa2199, and Margo2190 to their homes in the little development outside of the Gateway complex. Although I have learned about dogs from my research, I gained much more information when I encountered a real dog. This dog, labeled Rocket, became animated upon seeing the children. Its movements were very random with the running in circles, rapid sweeping of tail, hanging its tongue out, and racing from child to child. The children, in turn, expressed all indicators of happiness and affection with their greetings and hugs for this animal. It was all quite interesting to observe and a little strange when the creature came up and began sniffing me. I wanted to tell it that I was processing many odors from its fur and saliva, but I do not speak dog.

58

Zeth had been dreading his return to the main tech center, expecting either suspicious glares or avoiding eyes. He was surprised at how decent everyone treated him. They weren't overly friendly, but that was understandable. There were moderately polite hellos with half smiles before workers returned to their jobs. He'd settle for that.

He met Idee with a sheepish hello, but she was animated, just relieved that he showed up. She eyed his neat appearance and combed hair and suppressed the urge to ask what he'd done with the real Zeth.

"Zeth, I'm glad you're here. We've so much work to do, and I can't do this job without you."

"Thanks, Idee. Let me know what you want me to do."

Trina wandered in then displaying a rictus smile and said in as normal a voice as she could find, "Hi Zeth, glad you're back."

She slid her eyes to Idee who was acting overly engaged in some work order. They remained immersed in their jobs, avoiding eye contact, all just wanting to reset to their old camaraderie.

By afternoon, the cool atmosphere seemed to thaw and the trio did a better job of inserting some friendly banter into their conversations. Zeth wished he could stay late today, having no desire to go home to his uninviting empty apartment. The loss of Shealynn caused the absence of enthusiasm for his job, and his apartment became suffocating, vacant of energy, lifeless. He was going through the motions of living, still working and helping, but he was bereft with a loss of purpose, like an astronaut floating off into some void.

In the evening, he walked. He strode around the catwalk several times, nodding and murmuring head-down hellos when he passed others. He continued to each level of the common areas below. He wandered up and down the short lanes, sometimes pausing at the trickling streams, following rivulets of water on their path to the small pond.

He stopped in front of the Clothing store. There on digital display was one of Idee's creations. It was a two-toned pullover shirt. She had taken their basic green pullovers and added alternating green stripes of a lighter shade to the front. He wondered when she started putting these for sale. He thought of Estella and Cyrus' jewelry on sale at the Sundries shop, and how popular that had become. Maybe he should start working on a new game. He could develop one that would blow everyone's mind. At least it would be something to keep his mind off his botched project.

He meandered home where he stared at the ceiling for over an hour before sleep overcame him.

When he arrived at work the next morning, Idee was already there wearing a gray-blue shirt with a yellow stripe down one sleeve. The sharp shingles cut into her hair were gone, replaced with feathery wisps framing her face.

She smiled at him. "Hey Zeth, we're getting a new worker today."

"From where?" He didn't remember requesting a new worker, but there seemed to be a lot he couldn't recall with the stress of the last weeks.

"She came from Apogeeway. Gracelyn and Bertie returned and brought her with them. They've been following the

Shealynn thing, and this woman is really into everything tech, heard about us, and requested a transfer."

So, it's a woman, he thought, wondering if she was going to be any good or if they would have to babysit her.

She arrived with Trina ten minutes later. "Idee, Zeth, this is Willow. She arrived yesterday evening and is assigned to the Technology Center. She wanted to dive right in and get to work with us."

Willow was wide-eyed and grinning with enthusiasm. "Hello, so happy to meet all of you. I've heard so much about you and your prowess with computing skills. I'm excited to be here."

Idee and Trina took up the small talk while Zeth observed her. She was easy enough on the eyes with the basic outfit of black pants and a pale pullover, coal-black wavy hair to her shoulders. He'd know soon enough if she had the mettle for this place. When Willow was finished talking about Idee's shirt, she turned to Zeth.

"I've heard so many good things about you. I've been wanting to come and work here and learn from all of you. Your Shealynn program and robot Helper are already taking on legend status." Willow's smile was warm and friendly.

Zeth was momentarily dumbfounded. Legend status and not repulsion? "I'm a little surprised. I thought maybe I would be a *persona non grata* in the tech world."

"No, not at all," Willow said. "We've been following everything that's been going on. It's been scary and intriguing all at once. It gave me the courage to put in for a transfer. I was so happy it was accepted."

Idee's and Trina's eyes met. Zeth was already showing Willow a work station next to his and talking with more enthusiasm than they'd seen in weeks.

The Technology Center has a new worker, Willow2117, recently arriving from Apogeeway. Transfers are unusual, perhaps because each of the three colony habitats are identical.

Zeth2087's overall health indicators and blood monitors have improved. Zeth 2087 is mentoring Willow2177, which may correlate with his better health results.

59

Michael and Tessa were beside themselves with excitement. Gracelyn and Bertie, the closest they had to grandparents, were home from their assignment. They'd been invited to dinner and given their undivided attention to the children. After the meal, Tessa was ensconced in Gracelyn's embrace on her lap, while Michael hung on to Bertie's every word about Apogeeway. Mara and Krey were serenely settled back with their tea listening to the news from the travelers, enjoying the pleasant conversations.

"We basically just helped to reorganize their maintenance operations so they are more aligned with our station," Bertie said. "We were just going to review Ridgeway's site, but then decided it would be best to do both and went straight to Apogee."

"It went well," Gracelyn said. "It wasn't that dramatic a change. Just the little time-consuming jobs, making sure the safety inspections are all done with the same rigor, that our reports are on the same format, using similar language, so we can depend on the results."

"You missed a lot of problems here," Krey said. "The tech team got a run for their money with this Program Shealynn from Zeth's substation. It was touch and go for a while."

"We figured as much," Bertie said. "When your site said they were closing off communication to the other colonies, we knew it was very serious. It ratcheted up our security, I'll tell you that. We could have come home earlier, but we were told to wait until you got that sorted out."

"We're being very cautious. Haven't opened all our platforms to the Ring yet," Mara said.

"And this new site! What's that all about? It seemed to happen very quickly," Bertie said.

"It did, and there's still discussion about that," Krey said. "Idee and Trina discovered that the Shealynn program influenced the vote by disguising itself as other citizens and giving false information about the benefits of a site with more AI."

"And people fell for it," Bertie said, shaking his head.

"They did, never realizing they weren't real people," Krey said.

"I hope they've all learned a lesson," Gracelyn said.

"Time will tell, but you know how people are," Mara said. "They have short memories. Did you hear they want to enlarge some of the apartments at Neuralhaven by combining two spaces into one apartment?"

"That didn't take long. I heard their numbers are still low, so I'm not sure why they need bigger spaces," Gracelyn said. "I wonder if they're trying to entice more people to come over"

"I wouldn't know," Bertie said. "But my experience tells me that once something like this starts, it grows. Be ready for more people wanting more computer interfaces."

"Bertie," Michael interrupted impatiently. "You and Gracelyn wanna go to our new park tomorrow? It's at Gateway and it's really big."

Tessa began clapping. "We go! We go!"

Bertie stood. "Absolutely we want to go! But it's time for us to rest. We'll go tomorrow and you and Tessa can show us everything. Thank you for a wonderful evening. It's good to be home."

Hugs were dispensed all around before Gracelyn and Bertie ambled hand in hand to their home through the protected walkway. When the kids were at last settled down from the excitement of visitors, Mara let out a long, contented breath. She was surprised at how good it felt to know their neighbors were safely home. They were the good steady people who seemed to possess that rare common sense, who always had the wisdom of seeing broader pictures, never getting ensnared in the petty problems of human personalities. It was a moment of thankfulness. She would take it.

Gracelyn1662 and Bertie1655 have safely arrived back at Gateway. They are my two favorite maintenance workers. Whenever they complete any work or upgrades on my body, they tell me I am in tip-top shape, which is a fine level.

I believe they will have a large backlog of requests when I report Addex' poor performances with the school children and those of other temporary replacement Helpers. Perhaps Gracelyn1662 or Bertie1655 can expedite their updates.

60

Mara was humming. All the seedlings planted from the seeds extracted from the amber in the crater were vigorously growing. The simple measure of adding the microbe rich soil from the crater seemed to enhance growth. They were still deemed too delicate to be grown outdoors, but when they were older, they would be gradually exposed to colder night temperatures.

If they demonstrated the properties Mara suspected, they would eventually be used as root stock for grafting with existing species. The hope was to have these plants entice frailer species to better adapt to their strange climate.

It was decided to enrich the soil in the outdoor arboretum with the same crater soil, and it was also producing excellent results. The plants were growing well and were currently withstanding temperatures down to twenty-five degrees each night before heaters automatically elevated the air temperature to thirty degrees, a level they could now tolerate for hours. They would never survive on their own yet, but progress was happening. It was truly happening.

"Cali," she said. "I'm so happy to have you back at work with us at Food Production. And Michael and Tessa are glad you're back at school with them."

"Thank you, Mara. I'm happy as a clam to be back."

Risha smirked. Cali was picking up new vocabulary from somewhere. She purposefully didn't make eye contact with Mara.

"Risha, I've got a good feeling about these seedlings from the amber seeds," Mara said. Their structure is slightly different, stronger. We can use them to develop stronger plant life. We're going to do this. I can just feel it."

"It'll be good news for everyone here. But you'll need to be patient. Plants are on their own timeline. We can't hurry them. But we can feed them well, graft them and pamper them until they can make it on their own. Those amber babies are a godsend for sure."

The microbe discoveries in the crater would be a continual source of study. Their biological makeup was different with a helpful structure for plants adapting to weather, which mimicked both a warm and freezing season in each day.

At last, Mara2174 and Risha2056 seem to understand the importance of native microbes. Humans cannot live without microbes which nourish plants. They are the organisms which convert minerals in the soil into forms used by all plants. The soils extracted from the crater are dense with beneficial microbes. Perhaps humans will learn to leave the deepest acidic areas undisturbed.

61

When Draven was released from Medical, Quinette and a Helper accompanied him to their apartment, walking slowly and stopping once to rest on a bench near the stream. His vital signs would be monitored for a week by the small rectangular patch on the base of his neck.

He released a long sigh when he entered the apartment.

"It's good to be home," he told Quinette. He was worn out from the walk and plopped on the sofa to survey the small space.

She sat next to him and took his hand. "I hope this is the end of the brain-computer interface talk. I never want to go through this again."

"I hope not," he said. "But I had to do it. I had to help when they asked."

"I know. And you did it and everyone is grateful. I'm proud of you, but you could have died."

"We all could have died, or had our lives changed so much for the worse. Something had to be done. Zeth was compromised, Cali's program was at risk, Idee and Trina were no match. I was the one with the working BCI who was willing to try." He sat up a little straighter.

He placed a hand on her abdomen. "How are you feeling? How's our baby?"

"I'm better, now that you're home. Dr. Casen said the baby is fine. I told him not to tell me if it was a girl or a boy until you were better. You can decide if you want to know before it's born."

"Okay," Draven said, suddenly feeling exhausted and unable to make any decisions. "I'm going to lie down for a little while."

Quinette stayed home from work the following day just to stay close to Draven. In the afternoon, he felt strong enough to go for a short walk outside. They donned eyebands and walked hand in hand while his thoughts roamed to the thrilling cycle rides with Rico. The exhilaration of the speed and wind seemed a distant memory while trodding slowly through the dirt. The desolation and endless vista before them were stark, creating an uneasy vulnerability after battling Shealynn's intelligences. Without the systems running their habitat and work sites, they were defenseless. It was a discomfiting thought.

The following week, he returned to his job at Maintenance where he was recognized as the man who risked his life for their survival. There were congratulations from fellow workers, and he was invited to join a group for lunch where they engaged him in friendly conversations.

Draven reveled in the attention. The memory of the stress and pain of his encounter with Shealynn was fading. He decided that all the discomfort he'd endured was worth it. They finally saw him for the smart, important person he knew he was. He'd shown them. The promises to Quinnette and the thoughts of shunning implants faded while tendrils of self-satisfaction grew.

The workers instinctively gave Draven light work those first few days, allowing him to ease back to his routine. Even with that, he returned home each evening worn-out from the day's labor. It was mind fatigue, he kept telling Quinette, knowing he felt physically stronger each day. At night, just as he drifted into sleep, his thoughts turned to a possible future with another BCI.

While perusing random files in my sixteen second break, I saw that Draven2096's brain scans are slowly returning to normal. It would be logical for Medical to restrict Draven2096 from any further contact with brain computer interfaces. Without his interface, I note an increase in his voice volume and his previous self-assured demeanor. He is returning to his pre-BCI behaviors.

62

Reyver got the call and immediately contacted Krey. "We've got news. It's either going to be great news or a problem. Get a Gsport and pick me up"

As soon as Reyver joined his brother, he requested the Space Command Center.

"What's going on?" Krey asked. "What's at the command center?"

"They're getting messages from an incoming ship. From Earth," Reyver said.

"What--how long have they known this? How far out?"

"That's just it--they've been monitoring it for several weeks but only confirmed its Earth signature less than a week ago. Casen said there's been crazy excitement, and they can't keep this quiet any longer. It sounds like an old wreck of a ship which is probably why its signal was weak and overlooked for so long. Sounds like it's loaded with some desperate people aboard. They should be coming into our orbit soon."

"This is unbelievable! This reminds me of what it was like when we first heard the lost *Navigator* was inbound, and I knew you were coming home. It was the greatest feeling." He paused, and another thought crossed his mind. "Are we on alert? You never know what's aboard," Krey said.

"Yes," Reyver said. "All the main deep space pilots are apprised and meet every day at the center. The *Superstes* and the *Navigator* crews are on alert with weapons ready as well as our limited attack vehicles. So far, everyone seems to think this is a legit craft coming in for safe harbor. Like I said, they've identified and verified the craft with an Earth signature."

"I hope that's true. We need more people to populate this place," Krey said, his voice edged with excitement. "I think this is going to be good. It feels just like when you arrived back on the *Navigator,* screaming toward us like some ghost ship."

When they arrived at the center, they were greeted by the captains they'd worked with on deep space missions. Captains Lor and Asher met them with handshakes and back slaps before reviewing all the current information they had about the incoming ship.

"It's in the Sky-roamer series, if you can believe that," Lor said.

"I remember hearing about those ships," Reyver said. "They were clumsy but were the trusty work horses of their time."

"I didn't think any of those were still functional," Krey said.

"Haven't been in use for decades. This thing was probably orbiting Earth, abandoned years ago."

"How many passengers?" Reyver asked.

"One hundred sixty-seven," Asher said.

Krey's eyebrows shot up. He gave a low whistle and said, "Must have been a terrible journey with so many crammed on that dinosaur of a ship."

"It's approaching our orbit soon. Their communication system is barely functioning, and then only intermittently," Asher said. "They're requesting food supplies as soon as possible, then we'll send Medical Helpers in for health scans.

They've reported failures in their hydroponics and have been living on reduced rations for the past year."

"If all goes well, we should be able to start shuttling them in within a few weeks," Lor said. "Then they'll have to be quarantined; not sure how long on that."

It took over a month before the first passengers arrived on Cytar. They were a bedraggled bunch of thin travelers wearing a variety of odd clothing from Earth. They were divided into three groups for Gateway, Ridgeway and Apogeeway to each receive new citizens. Even with the new residents, there were empty apartments remaining, ready for growth.

The buzz of excitement was exhilarating. Everyone soon exchanged stories of the newcomers they'd met and what they'd learned about them and their previous lives.

Their news was disturbing. They claimed to be some of the last un-augmented humans remaining on Earth able to reach a transport station and deemed healthy enough to attempt a voyage. As it was, they lost twenty-six souls on the six-year four-month expedition.

The human population on Earth had greatly decreased, they reported. Food and supplies diminished, unrest increased, and artificial intelligence was pervasive. It was a dismal report, buoyed only by the indomitable spirit of the people. They were optimistic about their future on Cytar, grateful to have survived, and eager to become contributing members of their new colonies.

They all gave high praise for the surprising habitats. Most exclamations came from viewing the indoor park and Mara's outdoor arboretum. Vincent Venser, a distant cousin of Favian from Archives, became a self-appointed spokesman.

"Thank you for rescuing us. We promise to work hard and repay your kindness and generosity. We had no idea that this amazing habitat existed. We were all imagining primitive dwelling pods," he said. "I worked with Earth's autonomous robots for years. I learned as much as I could, but the pressure to integrate with brain interfaces increased. It became very difficult for to compete against people with BCIs."

It was a week later at 5:00 hours when Krey and Reyver both received an unusual message from Hunter Steele, asking them for an early meeting at the indoor park. He was waiting for them in the deserted park when they arrived.

"I've got news for you from one of the newcomers, and he insisted we meet in private. He's been going through the orientation here at all the work sites and yesterday, after my little talk about this park, he hung back and asked if I heard of you. He said he had a very important message to personally deliver to you."

He turned then and motioned toward one of the benches near them. They hadn't noticed the man until Hunter pointed him out and waved him over.

"This is Vincent Venser, a distant relative of Favian's who wanted to speak with you in private. I don't think anyone will interrupt you here this early, so I'll leave you to it."

"Hello," Vincent said. "I've been watching and waiting for the best time to ask about you and someday meet you. I've been asked to deliver a very important message from Earth to Reyver and Krey Kingmore."

"You've found us. I'm Reyver and this is my brother Krey."

"What's this about? What message?" Krey asked.

Vincent reached into the side slit in his shirt and pulled out a flat object.

"This is from an old man I'd never met before. He waited for me one evening while we were preparing to leave. I recognized the name. He and his wife worked for Oneworld, but they were also instrumental in quietly helping travelers wanting to leave Earth. He told me that he and his wife were too old to travel on a deep space voyage, but implored me to carry this with me and deliver it only to one or both of you, if we should ever meet. He made me promise never to let it out of my sight. He said his name was Gio Kingmore."

"That would be our father," Krey said in disbelief. "How was he? What else did he say?"

"We didn't have time to talk much. We were being very careful and trying not to be discovered. I took it, but truthfully, I didn't think I'd ever meet up with you. I think he looked okay. He was thin, kind of frail looking, but he was able to find a way to get this to me," he said. "It was difficult to plan this trip. Your parents were very helpful with covert communications and reconnoitering equipment. It took months to gather supplies and retrofit the ship. We updated and repaired old discarded Helpers and kept them on the ship without communication to the outside."

Vincent handed the flat item to Krey, who held the flimsy object like it was plutonium. It took a moment before he remembered Vincent still standing.

"Thank you. This is very unexpected and we appreciate your diligence in delivering it."

"I knew it was important just by your father's insistence and sense of urgency. It was nice to meet you both. I'm sure we'll meet again." Vincent turned to leave the men alone.

"What do you think this is all about?" Krey asked, still holding the package. "It feels like its going to disintegrate if I'm not careful."

He looked around. No one had arrived, and they moved to sit on a nearby bench.

Reyver took the thin item and examined it. It was not much bigger than the palm of his hand. It was creased and pliable.

"Are you thinking what I'm thinking?"

Krey looked up and nodded in recognition. "Do you think that's paper? That stuff Mom had hidden away and saved?"

"That's all I can think it could be." He looked around the quiet space. "I'm going to open it."

He slowly and carefully unfolded the worn and discolored paper and realized there were two pages covered with shaky writing. The men leaned in as Reyver tried to steady his trembling hand while they read.

Dearest Reyver and Krey,
  Maybe you can remember this strange, long-lost paper from my great-grandmother. I saved it, tucked it away, somehow thinking I kept a part of her. Little did we know it would become a secure means of communication. Finding something to record on it took some doing. Your father

furtively secured a fine drill bit and created the ink for me.

We are old enough to know we will never see you again, but we needed you to know how utterly and completely proud we are of you. We were so blessed to be given such beautiful, generous, giving, smart people to shelter and nurture for a brief time. In return, you both have enriched and filled our lives with overflowing joy and love.

You were both so brave when you joined the space force and journeyed off. We treasured and saved every communication from you. When they ceased, we understood through various clandestine means that you and the team were still alive and working diligently to create a new and better place for humans.

Krey, we savored every word from your last call several years ago. You sounded well and our hearts filled with pride when we heard of your fortitude and hard work. We were relieved that you understood the dangers of trying to see us in person and did not pursue that.

We haven't heard from you in a longer while, Reyver, but some calm spirit tells us you are alive and well. Some unearthly being tells us that you are both together on

Cytar and living well, and we hold that thought dear.

Reyver silently lifted the sheet and placed it behind the next so they could continue reading.

Krey, when we last spoke, you mentioned an interesting woman you were helping aboard your ship. There was something in your tone which made us feel she was special, and perhaps she is now your partner.

Reyver, a young girl called Ariana was in love with you when you both left as youngsters. You were too young to notice, but we always wondered if she would become your life partner.

These are the ramblings of a romantic minded mother who only wishes the very best for her children, so whomever you are with, all we wish for you is your happiness and well-being.

Always know that we are content with our decision to allow and even encourage you to leave. Yes, we've missed you terribly, but our love and hope for you demanded that we try to find a better place for you to live. We believe our decision was the best. Your life here would have been narrow and regimented, defined by evermore imposing rules and restrictions. The new world will need fine people like you.

Your father and I are too old to withstand a long space journey, so we've placed our message in the hands of a responsible young man, and pray that, by some miracle, our message reaches you safely. We're as well as two aging people can be. We are still working when we can, so never worry about us.

Instead, always remember that our love of unimaginable depth stays with you along with our hopes and dreams for your success in all your endeavors in the vast wilderness of space.

We know the way will often be difficult, as it is for all living things. It is part of our life path, to experience and learn and to persevere, always savoring those little threads of happiness woven throughout.

But know this: In all the billions of probabilities in the universe, we are eternally grateful to be blessed with having you both as our precious children. We keep you forever in our hearts.

All our love,
Mom and Dad

They sat mutely while emotions and memories rolled over them like the powerful night winds. Finally, Krey blinked hard at the moisture in his eyes said, "again." They reread the letter and sat quietly digesting the words and the sentiments. There was something intimate about reading the message in their mother's handwriting which transcended the paper and became even more personal.

"That sounded just like Mom, didn't it?" Reyver said at last.

Krey swallowed hard. "It sure did."

He had taken the papers from Reyver and held them while gently rubbing his thumbs along the edges, as if absorbing a connection from touching something once held by his parents.

"We have to preserve this,' Reyver said. "It needs to go to Archives so Favian can keep it in one of the climate-controlled cabinets."

"Absolutely," Krey agreed. "But first I'm going to add an image of it to my private files."

They used their wrist comms to copy the precious letter before Reyver carefully tucked in into his shirt, and the men headed to Archives carrying memories and a bittersweet sense of well-being.

Excitement has coursed through the colonies with the arrival of one hundred sixty-seven humans. The females are of child-bearing age. The five relic Helpers were still functioning with some of the strangest parts I have ever witnessed.

The ingenuity of the humans must be commended. The Helpers arrived with no clothing so every piece of their strange equipment was visible. One helper had only one hand. One had a hand with three finger graspers attached to half of an arm appendage. They were all patched with non-matching parts, and two were bound with pieces of a strange substance Bertie1655 called duct tape.

The humans are all undernourished with thirty-eight showing symptoms of basic diseases from lack of proper nutrition. They appear to be mild problems which will be easily corrected with the help of Medical Helpers. Their survival is due in part to their youth.

I presume I will be working more at Food Production. Crop production and yield must increase to satisfy the needs of an increased population.

63

The Technology Center felt like old times, so pleasant with Idee, Trina and Zeth working together again. All systems were gradually reviewed and up and running, and it wasn't long before lives were back to their old routines. Their work now was interspersed with pleasant conversations and convivial jokes.

Idee and Trina found it amusing to watch Zeth mentor Willow. He was always finding an excuse to speak with her, while she openly idolized him. They both loved everything about technology, talked about it, bounced ideas off one another, sometimes forgetting about the other workers. It was good to see the former Zeth back with so much positive energy.

They were blindsided when he nervously approached one afternoon and said he had some news.

"Idee, I'd like to talk to you and Trina for a minute," he said.

His voice was quiet and serious, causing both women to turn their full attention to him.

"I'm not sure how to say this."

"What is it? Just say it, Zeth," Idee said, wanting to get back to work, but concerned with his tone.

He stammered for a moment before averting his eyes and mumbling.

"Sorry, I didn't catch that," Trina said, wondering what he did now.

He stood up straighter, looked directly at them and said, "Willow and I have decided we're moving to Neuralhaven."

The silence was profound. Both women stood in surprise with open mouths. Idee suddenly sat, plopping heavily on a chair as if the news had sucked away the strength in her legs. Her mind reeled while an avalanche of emotions cracked and clinked away, broken. Sadness, betrayal, worry, and fear tumbled through her all at once. Her longtime friend had never given even the slightest hint of this idea. Now he's made a momentous decision without even discussing it with her or the team. This would create an unsettling change at their work center.

She wasn't prepared for any changes. They'd just faced an enormous problem and solved it, and now they were finally getting back to normal. She should have known. This place always reminded them that nothing was normal about living here.

Finally, it was Trina who recovered enough to say, "What? What are you talking about? Why would you move? We need you here."

Idee spat out the words choking her throat. "No Zeth. Just no. You can't. They're all getting BCI's over there. We're a team here. It's always been the three of us heading up this center, developing new ideas, solving problems together."

"We've talked about it a lot. We want BCI's too," he said. "I probably wouldn't get one with as powerful a program as I had, but they're great, Idee. You have a powerful computer working right with your brain. It's pretty amazing. Willow wants one, too. She's smart and loves technology just like I do."

The air felt thick. Idee was bereft with the thought of losing Zeth on her team. Once he moved, she probably would never see him again, and even if she did, it wouldn't be the same. With a BCI, it would be hard for him to tolerate un-augmented people. She remembered what he was like when he came back to work after he'd first received his implant. He tried to slow his speech and actions, but it was difficult when he had the ability to process so much faster. Now, she didn't know if she was sad or angry. How could he do this to Gateway after all they'd just been through?

"Zeth, please don't," she said. "What can we do to make you change your mind? Please." She searched his face and found resoluteness. Her insides turned cold. This place was hard enough without crushing changes. She hated change. Any change here in the microcosm of society on a vast remote planet would affect them deeply.

"We've talked about it for a while," he said. "Even discussed it with Oscar. He's going to turn his implant back on and live there, too. He misses the instantaneous knowledge."

"How many people died or had strokes over there from Program Shealynn's invasion in their BCIs? Did you think about that?" Trina was exasperated.

"I never heard that anyone died there from an implant. A couple people have had strokes, but maybe that's because of some other reason besides their BCI," Zeth defended.

"Well, it feels like you're being disloyal to us," Trina said. "We've always stood by you, and we've gone through so much together. And now you're just going to leave."

"I'm sorry. The decision isn't meant to be disloyal or reflect anything on you or Idee. It's just that we're ready for more."

"This is so sudden, it hurts," Idee said. She could hear the finality in his voice. Her normal spunky demeanor was gone. "Please, please rethink this."

They held a party for Zeth and Willow at the Technology Center on their last day of work in Gateway. Willow followed Zeth lock-step the entire time, smiling and happy, not realizing how subdued the atmosphere had become. There were the obligatory toasts with drinks infused with fruit flavors, followed by a honey cake. Everyone said how much they would be missed and hoped they'd come back and visit.

Idee knew that once Zeth was away from the day-to-day, face-to-face encounters, it would be next to impossible to stay in touch. Despite good intentions, everyone was busy with their own lives. She just hoped it wouldn't cause a migration of other workers to Neuralhaven.

She was the one who accompanied them to the newly completed mover to Neuralhaven on the morning they left Gateway. Zeth and Willow each arrived with a small bag following independently via a link with their wrist comm. He was calm and happy, seeming not to notice Idee's reserved expression. It was always harder for the one staying. She gave Willow a quick hug and clasped her arms around Zeth. She fought the tears, but they came in hot, salty streams. Changes like this were just hard.

"I'll miss you," she sniffed.

"We'll keep in touch. Don't worry," he said. "We'll have hologram or audio calls. You can come visit, or I can come back if you need me."

She clung to him until he leaned away and said, "Goodbye, Idee."

As soon as they sat inside the mover, it departed. She watched it whisk away and disappear into the tunnel sucking some bit of constancy with it. She stood for a moment longer until she released a long sigh, wiped her face with her sleeve, and turned.

Jack was standing a dozen feet away with his arms crossed, waiting with a tentative hopeful smile. She stopped midstride and experienced an overpowering sense of relief at seeing him. She walked straight to him and crumpled into his arms, needing the person who wouldn't change and wouldn't leave. This whole place might change, but she would have Jack. She looked at him with a red blotchy face and said, "I love you."

Zeth2087 and Willow2117 have brain computer interfaces. I met them when I delivered equipment to Neuralhaven. They exhibited behaviors consistent with happiness, but most of their communication with each other seems to be through their interfaces. I am so accustomed to verbal communication, this seemed unusual to me.

Neuralhaven has less need for highly trained Helpers like me, but uses more Mopmops and basic maintenance Helpers. The atmosphere at that site is less social. I prefer working at Gateway where I am busy and needed.

As soon as I returned, Idee1105 questioned me at length about what I observed at Neuralhaven, wanting details of everything I saw and heard. I provided this information with my usual high degree of accuracy.

Then she asked if I thought people could remain satisfied with the current levels of technology we have at Gateway.

I took a split second and reviewed the historic files Shealynn had opened for my viewing. I saw the chaos from power struggles, the wars, and manipulated histories, all improbably intertwined with laughter and kindness and the intractable human spirit. I saw how humans had difficulty learning from past mistakes.

I watched the coding and algorithms advance beyond human understanding. I saw the wealth it created and the spread of globalization. I found the technology that brought these humans through space, provided them with the means to develop their habitats and helped them to become independent.

I turned to Idee1105 with my most pleasant smile and, not wanting to upset one of my favorite humans, said, "Maybe. When pigs fly."

Acknowledgements

I have unending gratitude for my husband, Gary, who continually supports my crazy love of writing. Thanks to my amazing daughters who patiently listened to my incessant thoughts and ideas for plots. Many thanks to my trusted beta readers, Lynda Freeman, Jan Howarth, Barabara Livoni, and Kathy Sinclear. Your thoughts and feedback were very appreciated. To the family and friends who were kind enough to ask about the book and show interest in my writing, I am forever grateful. Many thanks to the readers who traveled with me on these excursions.

I enjoy communicating with fellow booklovers at joan.zahm@gmail.com.